A Natural History

Middle Age

DAVID BAINBRIDGE

Portobello
BOOKS

First published by Portobello Books 2012
Paperback edition published 2013

Portobello Books
12 Addison Avenue
London
W11 4QR

A CIP catalogue record for this book is available
from the British Library

9 8 7 6 5 4 3 2 1

ISBN 978 1 84627 268 4

www.portobellobooks.com

Text designed by Lindsay Nash

Typeset in Minion by Avon DataSet Ltd, Bidford on
Avon, Warwickshire

Printed and bound by CPI Group (UK) Ltd,
Croydon CR0 4YY

Contents

The majority of mortals ... complain bitterly of the spitefulness of Nature, because we are born for a brief span of life, because even this space that has been granted to us rushes by so speedily and so swiftly that all save a very few find life at an end just when they are getting ready to live.

Seneca, *On the Shortness of Life*, first century AD

Twenty or so vigorous years between the end of reproduction and the onset of significant senescence does require an explanation.

Blurton Jones, Hawkes and O'Connell,
American Journal of Human Biology, 2002

Introduction

A New Story of Middle Age

It was not supposed to be like this. I was supposed to buck the trend. At thirty-eight I had derided the numerical milestone towards which I was hurtling, laughed in the face of the new decade of life which was *supposed* to make me middle-aged. Even the name of this life-phase – 'middle age' – seemed alien, a term of abuse more than an actual phenomenon I was ever likely to experience. It sounded like something which affected one's parents rather than oneself. No one told me when it started, or ended, or what might happen, so I assumed it was all a hoax, a fable invented to force people to grow old before their time. I felt young at thirty-eight, and I still felt young on my fortieth birthday – not for me the clichés of middle age, surely? Yet now I am forty-two, and suddenly time is screaming past, contemporary popular music means nothing to me, and I have a belly, miscellaneous aches and a sports car. What on earth is happening?

Apart from the disgraceful fact that it is happening to me, the biggest problem with middle age is that it is very difficult to define. Often it seems we simply cannot make up our minds about middle age. Is it an annoyingly vague period of life or irritatingly distinct? Is it a state of mind or a pre-programmed biological life-stage? Maybe it begins when becoming eighty years old seems alarmingly close rather than eerily distant. Or perhaps it is the time when you face the choice between two competing alternatives: accepting that life is going to get inexorably worse, or steeling yourself for the constant self-deception that things are better than they have ever been. Most of all, I wonder why middle age strikes fear into so many of us.

In this book I want to find out what middle age *is* and what it is *for*. On our journey, we will try to define 'middle age' – that in itself is no easy matter. If you ask a doctor to define middle age, they will probably talk about the menopause. If you ask a sociologist, he may mention empty nests and tolerating teenagers. If you ask an economist, she will explain career-peaking, maternal return to work, and provision for old age. If you ask a friend, he might tell you it was the moment he looked into the mirror and realized he was turning into a replica of his parents. But do any of those things truly define middle age any more? After all, men do not have a menopause – well, not in the same way women do. Also, many middle-aged people now care for young children, adult children, new partners' children, or no children at all. In addition, many people's careers do not peak in middle age, and many mothers return to work a long time before middle age. In short, people do so many different things with their lives that none of these oft-suggested definitions seems to crystallize the phenomenon we are trying to explain.

So what is my solution? I am a reproductive biologist and a veterinary surgeon with a zoology degree, and I believe I can offer an approach to middle age that will sort out all this confusion.

For a long time I have been fascinated by what utterly bizarre animals we human beings are. From a zoological point of view, many of the things we do in our lives are absolutely freakish. We are the strangest creatures on earth, and we are strange for all sorts of reasons. In the past I have applied a zoological/natural history approach to examine the idiosyncrasies of human pregnancy, the genetics of human sexuality, the human brain, and even teenagers. The same approach can be equally enlightening when applied to that most subtle and complex of human innovations: middle age. Human middle age is a unique phenomenon – it is quite unlike the middle of any other animal's life. And in this book I will show you that the uniqueness of human middle age springs from the influences which made us evolve it in the first place. Being middle-aged is not about being old – it is something altogether different.

I hope to persuade you that standing back and viewing middle age as an exceptional characteristic of an exceptional species can tell you more about your own mid-life than you ever thought possible (whether your own mid-life is past, present or future). At the start of this book, I will provisionally define middle age as occurring between forty and sixty years old, and whenever I refer to 'middle age' without being more specific, you can assume that I mean those fifth and sixth decades of life. This range is, of course, entirely arbitrary and I have discovered that when I propose it, older people think my numbers are worryingly low and younger people think my numbers are strangely high. Yet setting these arbitrary limits demonstrates how much of our lives we each spend in middle age, and the bewildering variety of things my zoological approach to human middle age must explain.

The essence of middle age is best evoked by the questions that might cross the mind of a middle-aged human in the middle of the night. Am I become biologically worthless? Am I getting ill more? Am I ageing at the same rate as other people? What did I develop that complex personality for, exactly? Have I improved my circumstances since I was a child? Should I save for my children's inheritance or my own old age? Have I left it too late to have children? Am I unhappier than I used to be? Why are the social rules no longer *my* rules? Should I want to buy a motorbike and run off with a model? What do I do now the kids have left/are leaving/won't leave? Who is this person lying peacefully asleep next to me and why do we two not love as we used to?

This may seem a daunting list of questions to tackle, but my zoological, evolution-based approach can answer them. I do not believe that we are entirely the product of our genes, or that society and environment have no influence upon what we are; nor am I saying that we have no control over the course of our own individual lives. However, I do claim that middle age is a specific, explainable phenomenon which is *largely* the product of millions of years of human evolution. Middle age is not a twentieth-century cultural invention. It is first and foremost an ancient biological entity, and everything else springs from that.

So why do I insist that middle age exists as a separate phenomenon at all? Is it not simply the start of a gradual, uncontrolled decline towards old age and death? One aim of this book is to show you that, despite the problems we may face when defining middle age, it is a definite, discrete part of human life, unlike any other. It is not just a matter of getting old; this is because of what I call the 'distinctive-abrupt-unique' triad of middle age. The first is that middle age is *distinctive*. Again and again, throughout this book we will investigate aspects of mid-life which do not exist in young adulthood or old age – middle age is simply different. The second feature is that the changes of middle age are often *abrupt*. Things which happen in middle age often happen surprisingly quickly – certainly too fast to be part of some gradual, cumulative, senescent degeneration. The third feature of middle age is that it is *unique* to humans. Taken as a whole, we do not observe anything like it in other species, and this makes it even more special.

I will review a wide range of evidence which explains why we have evolved middle age. We will see that for most of our evolutionary history, life has not been as nasty, brutish and short as we are often led to believe. For much of our time on earth, many humans have lived beyond forty, and there are special reasons why natural selection still acts on humans at such advanced ages, long after it has given up on other animals. Indeed, over the millennia natural selection has moulded over-forty humans to be the distinctive creatures they are. They are not fading away, but are instead entering a new, specific phase of life in which their social, emotional, physical, sexual, mental world shifts once more. It shifts to a new state which we now know is a creation of our evolutionary heritage, and the sooner you realize this, the sooner you can begin to make sense of your own middle age. Thus this book is written for anyone who is, has been, or will be, middle-aged.

Yet although I claim this book is meant to be read by anyone, existing research into middle age can often seem biased towards just one subset of the human species. Most of the published evidence is drawn from studies in people who fit what might once have been called 'the norm': heterosexual long-term couples with children, who happen to live in the developed world. Of course, it is important for us to be able to extend our conclusions beyond this group, into the many members of the human population who do not fit this description, but there are simple, practical reasons why the study of middle age has become so skewed. Any scientific study works better if it includes a large and clearly defined sample of the population. If, for example, you want to study attitudes to child-rearing, or the effects of the menopause on people's sex lives, then it is simply *easier* to study a large homogenous group of people who are all likely to have children or have an ongoing sexual relationship – and long-term heterosexual couples represent just such a sample group, especially as they often conveniently badge their status by getting married.

Such justifications for the bias inherent in middle-age research are pragmatic, but there is also a very good evolutionary reason why we should be especially interested in breeding heterosexual couples. As we will see, a great deal of what constitutes middle age is determined by our genes, and the simple fact is that we have all inherited our genes from a long ancestry of child-raising heterosexual couples. This does not mean that homosexual people, or people who choose not to have children, or people who opt to live in non-couple relationships, are not important – they are, and we must bear them in mind throughout this book. However, it is essential to recognize that their middle-age changes, too, reflect the genetic inheritance of millennia of child-raising couples.

This book is divided into three parts. In the first part we will look at how we ended up with middle age in its present form.

We will explore the processes which created the human life-plan, and even look back to the fossil evidence for middle age. This will then help us to explain the many physical changes of middle age, and how the ageing process has been tempered and postponed in our one, lucky species.

The second part of the book will focus on the middle-aged mind, mainly because the brain constitutes so much of what makes us different from the beasts. We humans spend so much time just *thinking*, that a third of a book seems the least we should devote to how our intellect, personality, psychology and emotions change in mid-life. We will even pause to consider why time appears to pass so rapidly when we reach middle age.

The third part may initially seem to take us forward, into sex, romance (either surviving or wilting), menopause, babies, families and *that* crisis, yet it will soon become apparent that it is taking us back, too. It will return us to the great biological-evolutionary reckoning of middle age – the time in our lives when we humans consciously reassess what our life has been for, while there is yet time to change it. And at the end of the book we will pull all the threads together in an attempt to generate a coherent theory of what middle age is and what each of us should do with it.

The information on which this book is based has been drawn from a wide, and sometimes uneasily eclectic, variety of sources – mainly the sciences, but with a smattering of the arts and social sciences thrown in. As I hope will become clear, science is the best of these sources because scientists are honest about the limitations and failings of their methods. Yet as we will see, even science can have trouble mixing things like genes, bodies, emotions and cultures. Perhaps it should not surprise us that explaining middle age requires so many differing approaches. After all, it is a topic which has been wrestled with by biologists, philosophers, medics, historians, mathematicians, novelists, sociologists and even engineers.

Yet through all this, you should bear in mind that the news is good. Middle age is important because you spend a large chunk of your life in it, but you, dear reader, are fortunate enough to be alive at the best time and place to be middle-aged. It is not that people in the modern developed world never die young, but people who do die before the end of middle age might be justified in feeling pretty hard done-by. After all, when you are middle-aged, things are certainly changing, but you are not yet old. Or at least that is what I keep telling myself.

A strong thread running through my new story of middle age is that human mid-life is a positive, not a negative thing. It is flux, not crisis. I would even go so far as to say that it is a liberation – evolutionarily, culturally and personally. I hope to convince you that now is the time in human history for us to reclaim the healthy, productive middle age which was our ancient birthright before drudgery, contagion and filth stole it from us.

PART I

FROM SAVANNA TO SUBURBIA

Why Middle Age Has Never Been About Growing Old

I believe that our Heavenly Father invented man because he was disappointed in the monkey.

Mark Twain, autobiography, 1924

1. What makes middle-aged people?

To understand middle age, we must start with some basic questions. So many aspects of life converge to create the phenomenon we call middle age, that at first sight it may seem an insuperably large topic. In the fifth and sixth decades of our lives we change physically, intellectually, sexually, emotionally and socially in ways which are interwoven, complex and profound. Also, middle age differs dramatically between individuals, sexes and cultures. How are we ever to comprehend this bubbling ferment of change and variability?

We have to start somewhere, so I will start with a deceptively simple question: what processes lead to the creation of a middle-aged human? And as we will see in this first chapter, to answer this question we must understand three things: genes, development and evolution.

I

Genes come first.

Forming any sort of animal requires three things: energy, chemicals and information. In fact, making something as wonderful as a middle-aged person requires a very great deal of energy, chemicals and information. However, we are not going to worry about the energy and the chemicals – we acquire those by eating and breathing just like any other animal, so it is not they which make middle-aged humans special. Instead we are going to concentrate on the information, because that is the interesting bit. Only humans have the necessary information to make middle-aged people.

A surprisingly large amount of the information needed to make a middle-aged person is stored in our genes – not all the information, but most of it. In the central nucleus of almost every cell in your body are forty-six stringy chromosomes, each containing extremely long chain-like molecules called deoxyribonucleic acid, or DNA. DNA has several properties which make it very good for storing the information required to make an animal. First of all, it is inherently stable and resilient, and we also have mechanisms to repair it so that it lasts even longer. Second, its famous 'double-helix' structure means that one DNA molecule can untwine and reform into two DNA molecules pretty much identical to the original – and this is what happens when cells divide. Third, DNA can be deliberately cut and spliced back together, and this is what happens when two individuals are manufacturing eggs and sperm with a view to making babies.

The fourth and final useful property of DNA is that it can be used to *make* things. In each human cell there are roughly 23,000 stretches of DNA which can be put to creative use, and we call these stretches 'genes'. The DNA chains which constitute these genes are not uniform along their length; instead they are formed by the joining together of four different elements (called *A*, *C*, *G* and *T*) and those blocks may be threaded together in any order like red, green, blue and yellow beads on a string. For example, a gene which is used to make type I collagen (a protein which makes up much of the mass of the human body, helps stop middle-aged people sagging too much, and is often mentioned in adverts for anti-ageing cream), starts with its building blocks in the order: *ATG TTC AGC TTT GTG GAC CTC CGG CTC CTG . . .*

This may not look like a very promising start, but bear with me. This DNA 'genetic' sequence is only useful because there exists a complex machinery in every cell which knows that the sequence is a code. Special molecules continually decrypt this gene code to make molecules (usually proteins) which do everything to make the cell work. Genes can make all sorts of proteins – proteins that chop chemicals up or join them together, proteins that allow things to flow in and out of cells, proteins that cause movement, or proteins like collagen which provide structural support. In fact, almost everything that goes on inside your body is the result of the activity of the molecules produced by those 23,000 genes.

All these codes and products may sound arcane, but it is important to realize that these genetic codes are all the information that most animals ever get. If you hatch from an egg and wriggle away with no parental care at all, the only thing you have to steer you through life is the information contained in the genes you inherited from your parents. Those genes are the sum total of the guidance you are given to form yourself, grow, behave and breed. Without genes, we would be nothing.

As we will soon see, humans get additional information to that which is encoded in their genes, but *not much*. It is remarkable that only 23,000 coded instructions constitute most of what is required to generate, operate and maintain a middle-aged person. In fact, when the human genes were first totted up, biologists were shocked to discover that this number was so small. Many cars have more components than that, and they cannot do a fraction of the things a person can do. When you consider that roughly thirty of those genes are required to make all the different types of collagen, and over a thousand are set aside for detecting smells, you can see that there will be surprisingly few genes left to coordinate complex things like making babies and orchestrating a mid-life crisis.

II

So much for genes. Now for development.

Every so often I am asked to give talks about biology to engineers and architects at my university. There are many justifications I use to claim to them that 'my machines' of flesh and blood are much cleverer than 'their machines' of steel and glass, but usually the most convincing is that my biological machines have to develop and grow autonom-ously, all the while still functioning as a living organism. For example, there is no stage in human development when our constituent parts can lie scattered and idle on a work-bench, waiting for some benevolent constructor to put them together. Instead, humans and animals must self-assemble, and they must remain functional throughout this self-assembly.

In fact, this self-assembly is the most bewilderingly won-derful thing that our 23,000 genetic instructions coordinate. It seems such a demanding feat that we suspect it is the primary function of many of those 23,000. As a result, devel-opmental biology is a big part of modern biological science, with thousands of scientists around the world labouring to identify the processes involved as a single-celled fertilized egg converts itself into a large, fully functioning, complex adult animal. Once again, it turns out that the vast majority of the information – the instructions – required to make a mature human being is contained in the simple *A-C-G-T* code of the genes. Developing animals are awash with cascades of genetic activity, as individual gene products switch on other genes, which in turn fire up yet more genes. This array of gene products then produces hands, ears, kidneys and hearts by inducing cells to proliferate, migrate, cooperate, specialize or die in very complex configurations.

Modern developmental biology has shown us amazing things. For example, some genes are so useful in forming bodies that they have been used repeatedly throughout evolution. Because of this, many genes are common to the development of humans, mice, fish, flies and worms. It is as if we all share a common genetic toolbox with molecular spanners and protein screwdrivers which can be turned to almost any purpose. Indeed, many of these genes are so useful that they can be reused many times during the creation of a single body – for example, a single gene may be used to create things as diverse as brain, liver, bone and testicle. This reuse of the same genes for different purposes is probably how we can get away with having as few as 23,000 genes, but it also means that these multi-purpose genes must be used very carefully. Otherwise you might end up with testicles in your skull.

All these developmental discoveries have changed the way we think about ourselves, but there are two important points I would like to make about how genes control the construction of a human body. These points are irrelevant for most of developmental biology, but they are crucial for us because we are interested in middle age, and middle age is special.

First, we must not be misled by the fact that most developmental biology has focused on what happens before we are born. This focus is entirely understandable, because the science was driven by an urge to find out how something as spectacular as a baby can be forged from something as insignificant as a fertilized egg. However, development is certainly not only about embryos and fetuses: after we are born there is still a great deal of developing left to be done. This post-natal development is just as crucial and gene-driven as the pre-natal kind, even though it occurs at a more leisurely pace. For example, for the first two years of life the brain keeps growing at the same rate that it grew before birth. And, after a pause, the reproductive organs suddenly start a rapid process of development in the early teens. The limb bones also keep growing in fits and starts throughout the first two decades of life. Yet even that is not the end of development. One of the central ideas of this book is that the developmental programme does not stop at birth, or puberty, or skeletal maturity. The genetic 'clock of life' just keeps on ticking and people keep on changing far into adulthood. We will see that there is a positive, active series of genetic events which continues long enough to cause things like the menopause and middle-aged spread. A striking example of this is the distribution of male body hair, which changes and develops continuously throughout the first sixty years of life. Changes as specific and distinctive as this simply cannot be explained any other way, and certainly not as part of a process of uncontrolled deterioration and ageing. A middle-aged human must continue to develop in much the same way that a human fetus develops – otherwise middle-aged people would just look like tatty twenty-year-olds. You may be grown up by forty, but you have not finished developing.

The second point I want to emphasize about the development of middle-aged people is an unusual feature of the human brain. The brain plays a critical role because we are a very intelligent and very social species. However, it has one strange feature – it responds to change in other parts of the body in a manner quite unlike any other organ. Let us take social and sexual behaviour in middle-aged women as an example. Obviously middle-aged women think and behave differently from younger women, and we can expect this to be caused partly by genetic and cellular changes in the brain. However, there is another force affecting how such women behave: their brain is all too aware of the changes going on in their own body. Unlike other organs, the brain reacts and responds to its subjective perception of the body which it inhabits – it is self-aware. Whether you look young and beautiful or old and haggard will have a big effect on your self-image, attitudes and thought processes. People are all too aware of the importance of their place in the human social world, and because of this, their perceptions of themselves continually mould the way they think. Of course, whether you actually *are* old and haggard is itself largely controlled by age and genes, but in this context age and genes are not affecting your brain directly. Instead they are affecting the brain indirectly by changing the body in which that brain finds itself imprisoned.

III

There is one more thing we need to consider if we are to understand the origins of middle age: evolution.

The genes that make today's middle-aged humans are the genes we have inherited from our ancestors – generation after generation of humans, proto-humans and even pre-humans trying to make their way in the world, sometimes failing and sometimes succeeding. By the early eighteenth century many zoologists had concluded that animal species change over time, and sometimes even split into multiple descendant species. Because some species look so similar to others, it was hard to believe that each was formed in a separate act of divine creation. Instead, some animals just appeared to be subtly modified versions of others. This process of gradual change and multiplication of animal types over time was given a name – 'evolution', a term stolen from eighteenth-century linguisticians who had used it to explain how human languages have changed over time.

It was two British naturalists, Alfred Russel Wallace and Charles Darwin, who first proposed a convincing mechanism by which evolution might take place, and we now call their ingenious theory 'natural selection'. For centuries it had been noticed that animal populations vary – any species of animal contains large individuals, small individuals, fast individuals, slow individuals. And any decent stockbreeder knew that if individual animals were selectively bred, there would be a good chance that their offspring would inherit their parents' characteristics. In other words, animal populations could be seen as seething masses of variations which can by some physical (that is, not spiritual) means be passed down the generations. At the time, no one fully understood what these physical means were, but we now know them as genes.

Wallace and Darwin realized that heritable traits provide a means by which animal species can change over time and gathered a great deal of evidence to support their theory. The exact process of natural selection is very important for understanding how humans evolved middle age. According to natural selection, if certain traits are beneficial to an animal and help it to produce many successful offspring, then the genes which produce those traits will survive and be propagated in future. Over the generations this process happens again and again, with genes which promote successful reproduction accumulating while genes which do not promote it are lost for ever. As a species' environment changes, so the traits required to help it breed in that environment change too, with the miraculous result that the animals themselves slowly alter over the millennia. Thus, animals adapt to their changing environment, and this goes a long way towards explaining how evolution occurs.

This book is entirely based on the premise that human middle age results from millions of years of evolution of genes. Because of this, there are some issues which I should address straight away.

The first is the question of evidence: how much proof do we have that evolution by natural selection actually takes place? One line of evidence is copious, and this is the observation that the animals alive today and the fossils we dig out of the ground certainly *look* like products of extremely long periods of evolution. However, some would say that this after-the-event evidence is not good enough, so biologists have tried to observe evolution actually taking place. Because evolution happens quite slowly this is not an easy thing to do, but it is possible. Scientists have watched rapidly evolving organisms such as microbes change over the generations, and they have also observed evolutionary change in larger animals when they are placed under strong natural selection pressures – such as lizards being introduced to new islands. It has even been possible to observe evolution in stressed human populations, as in the rapid development of community-wide genetic resistance to the devastating spongiform brain disease kuru ('laughing death') in cannibalistic New Guinea tribes. All in all, the evidence for the theory of evolution by natural selection is pretty good, and human evolution is not an exception.

The second issue is an element of evolutionary science which has caused a great deal of controversy: the evolution of our psychology and behaviour. The way we think is such an important part of our lives that some scientists proposed that our psychology has evolved to its present state in exactly the same way that our physical attributes have evolved. Thus was an entire new field of science born: evolutionary psychology. Evolutionary psychology has its critics but I should come clean and say that I am not one of them: I consider it to be a reasonable approach to the origins of human behaviour – and certainly not just a way of confabulating 'just-so' stories of why humans ended up behaving the way they do. After all, why should genes which make us *think* in ways which promote our success not also be bred into us by natural selection?

The third and final thing to mention about the genetic evolution of middle age is a paradox relating to the central importance of breeding in the theory of natural selection. Put simply, if natural selection involves the propagation of genes which help us breed, then where does this leave people who are no longer breeding? For example, does Darwinism mean that only children and young adults are naturally selected? Are post-reproductive, middle-aged people evolutionarily irrelevant? As we will see, this is an extremely important issue for us to confront in this book. If people over forty are not subject to natural selection, then this would mean they did not really evolve at all. And this would be extremely awkward because evolution of middle-aged people is the whole point of this book.

I hope this brief tour of the basic genetic processes which created middle-aged humans, and which still create them today, has given you some hints as to why they are so exceptional. But this glimpse into our evolutionary origins leaves us with yet more big questions to answer. When in human history did middle age – and the panoply of quirky wonders it bestows on us – actually appear? Why does human middle age evolve if it takes place after most of us have finished breeding? And if genes are just *most* of the information required to make a middle-aged person, then where does the rest of the information come from? Our quest for a new story of middle age has only just begun.

2. What breaks middle-aged people?

It would be reassuring to think of everyone's middle age as being entirely the result of a carefully coordinated developmental programme, honed over the aeons. However, we all know there is more to middle age than that – something more sinister. Even someone as unremittingly upbeat as me has to admit that middle age is not an *entirely* positive experience. It has a downside, and that downside is ageing.

Life is not all about growing up and getting better. At some stage we start to get old and things start to get worse, and most people feel that this begins at some point in middle age. Biologists argue about the point at which the human body starts to age (possible answers include: middle age, young adulthood, puberty, birth, and even conception) but elements of the process certainly become evident in new and more striking ways during middle age. Although I do not believe that middle age is unmitigated deterioration, there are clearly some unmissable signs that we are not as youthful as we once were – hair goes grey, skin loses elasticity, and you start wishing that someone would turn the music down rather than up. In fact, one of the reasons middle age is so fascinating is because the developmental 'clock of life' that has controlled our development since birth is now superimposed on, and in conflict with, a second process of deterioration and senescence. Whether we should call this second process a 'clock of death' is debatable, but middle age is certainly the time of our life when those two processes – formation and deterioration – coexist in the most overt and unsettling way.

If we are to comprehend middle age, we must understand ageing – why we age, how we age and why we eventually die. Of course, in developed countries not many people die during middle age (probably only eight or ten per cent of people die between forty and sixty), so it is not actually death we are interested in here, but rather the nature of the ageing process which becomes so apparent during this period of life. One sign that ageing really kicks in between forty and sixty is neatly demonstrated by the fact that middle-aged people spend a great deal of their time talking about bodily deterioration (and in a later chapter we will consider why women find this especially alarming). Middle-aged people also think about death much more than younger people, even though they are not much more likely to die in the near future (in fact, they are usually less likely). And yet you rarely hear anyone over the age of forty expounding the virtues of the developmental programme which is driving them forward to bigger and better things. In order to appreciate the value of the changes we undergo during this period of life, we need to examine why we age.

So, we do not usually feel like we are ageing before we hit forty, yet we certainly do afterwards. Why, then, do we age?

The earliest ideas about ageing were very practical. For centuries it was thought that animals and people just 'wear out', like an old machine. There was thought to be nothing inherently 'wrong' with old people, other than that they were being ground out of existence by the stresses of life. This is actually a surprisingly modern-sounding theory because it considers animals and people to be rather like man-made machines with a limited ability to self-heal, and indeed large parts of contemporary gerontology are based on just this assumption. For example, it has recently been proposed that one major strain on animals is simply carrying out their own metabolic chemical processes, and that it is the waste and heat produced by metabolism which eventually destroy us. According to this theory, animals with a higher metabolic rate should die earlier – 'live fast, die young', just like the bioenergetic equivalent of a sixties' pop star.

Yet there are problems with this simple view of animals and people just wearing out. For example, according to this theory, if you take a wild animal and cosset it in a comfortable and plentiful artificial environment, then it should live almost for ever. In reality, if you pamper a wild animal in this way it will indeed live longer, but not *that* much longer. For example, chimpanzees in captivity may live twice as long as they do in the wild, but that is not really a spectacular increase if 'wear and tear' is the all-important cause of death. Indeed, animals living in comfy captivity often live longer simply because they are not killed by predators, rather than because they do not age.

Even the more modern versions of this theory have not stood up to scrutiny. For example, although it has been suggested that having a high metabolic rate may reduce an animal's life expectancy, surveys across many different mammalian species have shown only weak correlations between longevity and metabolism. Similarly, despite the fact that decreasing metabolic rate by reducing food intake may sometimes increase life expectancy, the effects are inconsistent and variable between species.

Because scientists began to suspect that ageing is more complex than the body simply wearing out, they wondered whether there are any members of the animal kingdom which could provide a new perspective on ageing and death. And indeed they found them, because some animals are immortal. You may be surprised to hear that such creatures exist, but organisms which reproduce by binary fission – just splitting in two – are effectively immortal. An old amoeba is indistinguishable from a young one – it must maintain its internal machinery in tip-top condition in anticipation of the time when it splits to make two perfectly functioning off-spring. Although the situation is slightly more complicated than I have suggested, essentially, amoebas do not deterio-rate. Age shall not weary them, nor the years condemn. They just stay perfect.

However, there does seem to be a price to pay for immor-tality. First of all, we suspect that permanently maintaining your internal processes in perfect condition is very demand-ing. If an amoeba's DNA is damaged or a fault develops in its protein-manufacturing machinery, then it is essential that the malfunction is repaired quickly before it does any more damage. If similar problems develop in one of the cells in your body, then it is likely that you will make only half-hearted attempts to do anything about it. If a human cell goes wrong then it is not too much of a problem, whereas if you are an amoeba, that one cell is all you have and you must make great efforts to remedy the situation. Thus, immortality is extremely demanding. Also, immortal animals tend to be asexual, and asexual reproduction carries with it tremendous genetic disadvantages. So sex and death go hand in hand – we humans can allow our bodies to deteriorate because we can make pristine new babies which will grow up and replace us when we die. Admittedly, this is not the most comforting thing to write in a supposedly positive book about middle age.

Once the theory of evolution by natural selection came along in the nineteenth century, scientists developed a new perspective on ageing and death, and they started to view it in a new, more positive light. Indeed, they started to wonder whether animals actually evolved death for good reason: whether it might actually promote the propagation of animals' genes. It was proposed that senescence evolved because living an excessively long time wastes resources that would otherwise be available to the next generation. This may sound reasonable, but there are problems with this idea of ageing as a positive, beneficial process. First of all, most wild animals probably do not 'die of old age', but instead succumb to illness, accidents or predation. Because of this, evolving ageing should confer little advantage on the species. Second, many evolutionary biologists worry that natural selection does not really work like this. Natural selection is usually considered to act on individuals – by dint of their own reproductive success or failure – yet these 'altruistic death' theories imply that ageing and death benefit the next generation as a whole. (If there was some way that individuals could, by dying, selectively benefit their own offspring who carry their own genes, then the theory would make more sense.)

Really, these problems expose the central contradiction of evolution, ageing and death. If evolution is all about the propagation of genes which cause beneficial effects in individuals, then how can things like ageing and death, which destroy individuals, evolve? Putting it crudely, evolution tends to make animals 'better' and ageing makes them worse – the two seem incompatible.

As a result, scientists who study ageing have divided into two factions. One faction believes that ageing evolved as an *active* process, a feature built into animals to make them deteriorate and die, thereby benefiting their offspring in some way. The other group believes that ageing did not evolve for its own sake, but that it appeared *passively* as a side effect of other processes, such as the fact that it makes more sense for an animal to concentrate on breeding than maintaining its own body. According to this theory, there is no programmed ageing – no 'clock of death' – but just gradual failure of the body because other things are more important than repairing it.

This active-versus-passive dichotomy pervades the contemporary study of ageing, and it is of great importance for people who are coming face to face with signs of their own ageing process – literally so, as they gaze into the mirror at their wrinkles and paunch. From the selfish point of view, if we want to ameliorate some of the unpleasant aspects of middle age, we need to know what ageing *is*. For example, if ageing is an active, specific, programmed biological mechanism, then presumably we could delay ageing by attacking that one single mechanism. Alternatively, if it is a cumulative process of widespread failure throughout the body, then all we can do is tackle each little failure as it occurs.

In the middle of the twentieth century, the passive theory of ageing started to predominate. The initial approach was once again to think of the human body as a machine – a complex machine with multiple parts, each of which is prone to break down. As time goes on, more and more of those parts stop working. When we are young, so little of our body has failed that we do not even notice it, but gradually the damage becomes evident. As the years pass more and more parts of the body cease to work and these many imperceptible little failures eventually accumulate to become the symptoms of ageing and lead inexorably towards death – the 'multi-hit' theory.

In addition, the new theories of ageing provided reasons why animals do not put more effort into alleviating the effects of ageing. This is because, even in very long-lived species, young adults are destined to contribute more genes to the next generation than old adults. An old human may age well, but they can never have as many procreative years ahead of them as a young human. And because breeding is all-important in natural selection, this means that natural selection acts predominantly on the young – the young will make more babies, so it is the young who evolve. The old are still subject to natural selection, but not as much – so the later stages of life do evolve, but there is less pressure on them to do so. This is yet another reason why middle-aged people are interesting: they are not evolutionarily crucial like young people but they are not evolutionarily irrelevant like the elderly. Instead they are in a fascinating grey area in between.

One result of this reduced evolutionary importance of older adults is that the human species tended to accumulate genes which adversely affect the elderly. This is not a malicious process – we have not deliberately created genes to make older people's lives difficult. Instead, genes which get damaged or altered so that they cause deterioration in later life simply are not removed from the population. Over the course of human evolution, individuals have bred well despite carrying these genes because they do not damage them until they are much older. In this way, the human race never lost the genes which cause the gradual degeneration of age – and because of this, we can all look forward to several decades of wrinkles, aches and pains.

There is another possible result of the passive theory of ageing, and this *does* sound rather malicious. This proposed mechanism of ageing has the awful name 'antagonistic pleiotropy' and it puts older people in a very difficult position indeed. According to this theory, genes which promote breeding in the young will be perpetuated, *even if* they cause adverse effects in the old. In other words, breeding is so important that all sorts of untold damage may result in the post-breeding years. Indeed, ageing could be seen as a result of the conflict between the requirement to stay alive and the undeniable fact that it is the young who breed. For example, sex hormones drive reproduction in the young yet promote tumour growth in the old. In dogs, for example, it has been found that castration reduces the incidence of tumours, and even reduces DNA damage in the brain in later life. The compromises are also very clear in mute swans, in that those which commence breeding at a younger age also *cease* producing cygnets at a younger age – they trade early-life fertility for later-life deterioration. However, despite all the evidence in animals, studies in humans are equivocal – for example, a retrospective study of longevity and fertility in European aristocratic (and thus well-documented) families failed to show that youthful fertility and longevity are incompatible.

An adjunct to these passive ideas of ageing is the 'disposable soma' theory which, unfortunately, just seems to make life even worse for the middle-aged. According to this theory, not only does natural selection fail to support older individuals because of the importance of the young, but it also selec-tively neglects the maintenance of the middle-aged body. Everyone knows that it is only our sperm and eggs which contribute to the next generation, yet few of us like to accept the evolutionary implication of this, which is that the only function of the rest of our body is to get those eggs and sperm together and make babies out of them. Thus our body (soma) is effectively disposable. Natural selection promotes maintenance and repair of the body only in so far as it pro-motes successful breeding, and surprisingly early in life (at twenty years old, perhaps), maintenance of the body starts to slip down our list of priorities. Even worse, your body is allowed to deteriorate somewhat in advance of actually becoming redundant – after all, you can drive your car without maintenance for a few years before it eventually breaks down. So when you are middle-aged, your ever-more troublesome body is already anticipating the negligible con-tribution you will make to the next generation when you are elderly.

Despite the dominance of these passive theories of ageing, there has recently been a resurgence of the active theories – theories that we evolved a specific programme of ageing as an advantageous process, a 'clock of death'. In short, the propo-nents of these theories argue that, although they may not fit with current theories of evolution, they actually coincide much better with what we observe in nature than passive ageing theories do. For example, if animals do not have active, built-in ageing mechanisms, then why do most species of animal have such characteristic lifespans? Hamsters usually start to look frayed by twenty-four months of age, domestic cats with good road sense often live to eighteen, healthy elephants usually keel over between sixty and seventy. In a conducive environment most people live to an age between sixty and ninety, which seems a suspiciously narrow range – man-made machines, in comparison, exhibit a far higher variability in their useful lifespan. It is striking to ponder why, at a time when many thousands of humans reach the age of one hundred, not a single one lives to more than one hundred and twenty-five. Our close relatives, the chimpanzees, almost never survive into even the sixty-to-ninety age range, suggesting that the alarm on their 'clock of death' is set to a different time. Indeed, the disparities between closely related species of other animals are even more striking. Surely there must be some active mechanism to explain why bats usually live five times longer than similarly sized rodents do, or why one species of mollusc can live *four hundred* times longer than another, closely related and similarly sized, species?

Further evidence for the 'clock of death' comes from species in which it stands out because it behaves in an especially striking way. In some, it seems almost cruel: for example, salmon and octopuses breed once and then immediately age and die extremely rapidly, as if a 'death switch' has been flicked. Conversely, a few animals show a striking variability in their 'clock of death', and as a result have lifespans that can vary by a factor of twenty within a single species – these are, perhaps, the exceptions which prove the rule. Others may have lost the clock altogether. For example, there is evidence that female painted turtles simply do not age – instead, their fertility and chances of survival seem to *increase* as they get older.

All these animals with unusual or non-existent active clocks of death serve to show that there may be more to ageing than just passive, uncontrolled degeneration. I am not sure if this is reassuring or not from the point of view of middle-aged humans, but it certainly helps to put the ageing process in context. First of all, it suggests that humans are pre-programmed to live at least as long as sixty years – and in the next chapter we will consider whether this actually occurred often during human history. And second, it allows us to think about why, within the constraints of the 'clock of death', middle age is such a different experience for different people – and whether there is anything each of us can do to ensure that we are one of the lucky ones who appear to age more slowly.

Recently, a whole field of research into ageing has developed in an attempt to discover what actually happens to bodies as they age. Much of the focus has been on the elderly and extending the lifespan, but it is in middle age that many of the signs of ageing first become apparent.

Because of the importance of the DNA genes in coordinating the functioning of our bodies, many scientists have directed their attention to the role of DNA damage in ageing. After all, if chemicals, radiation or cosmic rays damage enough of the DNA in a cell, then that cell simply will not work any more. Levels of DNA damage do seem to increase with age in humans and other animals, with species which possess more efficient DNA repair mechanisms achieving greater longevity. Presumably, we humans have become especially good at this, as we now live longer than almost all other mammals, with the exception of some whales.

There is also evidence that the structure of our DNA chromosomes is itself important in determining our lifespan. The end of each chromosome is capped by a repetitive sequence of As, Cs, Gs or Ts called a telomere. In babies the telomeres are very long but as we grow and our cells divide, chunks are removed from the telomeres and they become shorter and shorter. Once a cell's telomeres are truncated below a certain length, then that cell can no longer divide. In this way, telomere shortening places a limit on how many times our cells can divide. Experiments have shown that human cells grown under artificial conditions shorten their telomeres until they can no longer divide, but that artificially preventing telomere shortening allows cells to go on dividing for much longer. If every chromosome carries within it its own little telomeric 'clock of death', this may be one pre-programmed reason why people age. There are problems with the theory of telomere shortening, however, in that not all cells actually divide in adult humans. So, while telomere shortening could explain the deterioration of bone marrow, gut and testicle, where cell division continues throughout adult life, it cannot explain degeneration of brain, muscles and bones, where cell division almost ceases in adulthood.

In fact, rather than providing us with our elusive 'clock of death', there is some evidence that telomere shortening may instead have evolved to prolong life, by protecting us against cancer. Telomere shortening may arrest the uncontrolled proliferation of tumour cells – once a tumour reaches a certain size, its stunted telomeres may simply stop it in its tracks. And indeed, modern imaging techniques now suggest that many of us regularly develop tumours which grow to appreciable size but which then 'mysteriously' regress. If this tumour-limiting mechanism is indeed the reason why telomere shortening evolved, then it could be argued that ageing is a paradoxical side effect of a system which stops us dying from cancer.

Another theory of what causes ageing involves 'reactive oxygen species'. Many metabolic processes inside cells produce oxygen compounds with incomplete chemical bonds, and these wreak havoc with other molecules in the cell, damaging proteins, membranes and even DNA. Although our cells contain antioxidants to defuse reactive oxygen species (this is why you need to eat vitamin E), some of them still evade our defences and damage our internal cellular machinery. This continuing chemical damage could be a real problem in cells which do not divide and replenish themselves (brain, bone and muscle) and so it has been proposed that this is what underlies the gradual failure of human tissues during ageing. This theory has evidence to support it – some of our molecules have been shown to become more damaged as we age. However, it is unlikely to explain ageing entirely – for example, no one can explain why eating more antioxidants does not seem to make us live much longer.

The hunt for a genetic basis of ageing has heated up in recent years, with scientists desperate to identify individual genes which hasten or delay ageing. It is a commonly observed phenomenon that longevity is hereditary – long-lived people tend to have long-lived parents and some species may also be artificially bred to live longer. Experiments that have created Methuselah-like strains of flies and mice are intriguing as these strains have been observed to breed more slowly (which sounds like antagonistic pleiotropy) and to make more antioxidants (which suggests that reactive oxygen species may, after all, have a role in ageing).

The search for individual genes involved in ageing has also been given further impetus by our new understanding of a group of diseases which cause *premature* ageing. In these 'progeria' conditions the onset of ageing deterioration is hastened and often starts in early childhood, with many sufferers dying of 'old age' in their teens. Many of these diseases have been shown to be caused by damage to individual genes involved in DNA repair. As scientists have peered more closely into the workings of the ageing cell, they have found that genes implicated in ageing tend to fall into just a few groups, involved in the general maintenance and operation of cells – a rag-bag group including insulin and related molecules, 'deacetylases', 'NF-kB', the 'mitochondrial electron transport chain' and 'heat shock proteins'. And remarkably, there is now good evidence that the same sorts of genes may be involved in ageing in organisms as diverse as humans, mice, flies, roundworms and yeast.

As you can see, we do not yet possess a complete understanding of the genetics of ageing – certainly not to the point where we can use our knowledge to delay it. The main reason we do not yet have a therapy to extend life is that there does not appear to be a single master gene which controls life expectancy; instead, your lifespan may be dictated by the cumulative effects of multiple genes. Altering the function of even just one minor gene in an organism as complex as a mammal can have many unexpected and unwanted effects, so we must be extremely cautious about the prospect of manipulating the key genes involved in human ageing.

So middle age is a time of conflict, when your development 'clock of life' starts to be overwhelmed by the deterioration of age. Yet as we have seen, whether we should call this latter process a pre-programmed 'clock of death' or just a passive slide into senescence is a moot point. We do not yet know exactly why or how evolution made us mortal, ageing beings, but it is likely that delving deeper into our genes will provide some of the answers.

One thing we can be sure of is that life starts with development and ends with ageing, and that middle age is the special stage when the two clash most noticeably. Indeed, we might be tempted to provisionally define middle age as the time when the forces of creation and destruction compete on an equal footing. We are getting close to a theory of what middle age is, but first we must spend a moment peering into the distant past of our species. If we are to understand greying hair and crow's feet, we need to find out what middle-aged people were like when they were wild.

3. Are people really meant to die at forty?

> *Whatsoever therefore is consequent to a time of warre, where every man is enemy to every man; the same is consequent to the time, wherein men live without other security, than what their own strength, and their own invention shall furnish them withall. In such condition . . . the life of man, solitary, poore, nasty, brutish, and short.*

Thomas Hobbes, Leviathan, *The Matter, Forme and Power of a Common Wealth Ecclesiasticall and Civil*, 1651

Studying the anthropological history of middle age is fascinating. It can tell us where mid-lifers lived, with whom they lived, what they ate, and what ailed and killed them. If we want to understand modern-day middle age, then it is just plain interesting to know what middle age was like throughout our evolutionary past, and how many ancient people actually managed to survive as far as middle age. However, there is another important reason why we need to consider the ancient history of middle age, and that reason lies at the heart of this book.

Again and again I will argue that middle age is a special, novel part of the human life-plan which has evolved because it has benefits for each of us as individuals. However, for middle age to have evolved 'for good reason' in this way, it is essential for many humans to have survived to middle age during our evolutionary history. Put bluntly, if all prehistoric humans died at thirty-five, then natural selection could not act on the process of middle age and middle age would not evolve any beneficial characteristics. Instead, the processes of middle age which we observe in modern humans would be nothing more than a weird, abnormal symptom of suddenly being able to live far longer than we ever used to – than we ever were 'meant' to.

So how long did people live in the past?

We have good records of human lifespan over the last century. Within that short time, life expectancy at birth has increased dramatically in developed countries, to a stage when it is likely that at least eighty per cent of babies born today will live to the age of sixty-five. Most of the former causes of death – diarrhoea, pneumonia, septicaemia, wound infections, death during childbirth – have largely vanished, and we now die of different diseases at a greater age. This is an astoundingly large and rapid change that has resulted from improved nutrition, living conditions and medical care, and we can safely assume that it is unlike anything that has ever happened before in human history.

If we work backwards through the last few centuries, life expectancy gradually decreases to a point at which it may have crept down below thirty in the medieval period. And in the few thousand years before that, it is likely to have been much the same level. Yet even in those dark days of low human longevity, it is important to emphasize that things may not have been as bleak as they seem – if you were an adult, at least. Average life expectancy is a problematic measure of lifespan because it is enormously skewed by infant mortality, which was high until the last hundred years or so. Using the average life span of all humans who survive being born as a measure of longevity can be very misleading if large numbers of children die before they are five. In fact, if many children die before maturity, then the surviving adults have to live several more decades even to drag the average lifespan up to as paltry a level as thirty years. This skewed picture of low 'life expectancy' may explain why all through history there are reports of large numbers of people surviving well into middle age and far beyond – and not only the wealthy and privileged. Instead of looking at the population average, therefore, we should focus more on how many people who, having reached adulthood, subsequently lived past forty to reach middle age.

However, if we look back over the last ten thousand years, the story does not change a great deal. During that time the world was populated by country dwellers scraping a living by growing a limited array of crops, and town dwellers wallowing in the filth of unplanned, open-sewered settlements. As a result, it is unlikely that lifespans for most of the human population would be much greater between 8000 BC and 500 AD than they were in the medieval period. I admit that this is starting to look slightly worrying for my theory that middle age evolved because natural selection acted on a copious supply of ancient middle-aged humans. We have already looked ten thousand years into the past and still there is scant evidence of that copious supply.

Yet maybe things are different if we go more than ten thousand years into the past. Around that time, humans underwent the most profound change ever to occur in their way of life. The coming of agriculture was a remarkable phenomenon. It not only changed for ever the foods we ate, but it also compelled humans to settle down in one place, near their fields and livestock, and near where food could be stored. An agricultural way of life put a greater emphasis on property – ownership of land, farming implements and produce – and with it came the desire to defend that property from others and bequeath it to offspring. It is difficult to exaggerate the effect agriculture had on human existence. Some anthropologists think that entire populations switched from hunter-gathering to farming and settlements within just one millennium, and cities, writing and governments seem to have followed soon after. Even more remarkably, there is archaeological evidence to show that this transition to farming and settlement occurred independently in perhaps seven or ten different locations around the world at more or less the same time, possibly in response to climate change.

So agriculture arrived suddenly and it changed everything. Because of it, the last ten thousand years of evolution of *Homo* have been utterly different from the previous two million. In other words, 99.5 per cent of all human history – the time when our ancestors were approximately as big and clever as us – occurred *before* the advent of farming. Surely then, if we want to understand the evolution of middle age we need to look back to this pre-agricultural, pre-historic time – when most of that evolution occurred.

We often think of the lives of prehistoric peoples as being desperately harsh: 'nasty, brutish and short'. We might imagine that most prehistoric hunter-gatherer humans were stressed, starving and diseased, and only survived by the skin of their worn and chipped teeth. If this is true and very few people survived into middle age, my argument for an actively evolved, beneficial middle age collapses. So, how long *did* prehistoric hunter-gatherers live?

And with the word 'prehistoric' we come to the greatest stumbling block in our story. As a biologist dabbling in history and archaeology to study the story of middle age, I have come to appreciate the full meaning of the word 'prehistory'. History is what happens when people write down their experiences, and as soon as we have written records, we have access to ancient people telling us about how they lived and how long they lived. Before the coming of agriculture there is no history, no written record, so studying how people lived and how long they lived is tricky.

In the absence of recorded history, scientists have tried to measure the age of our prehistoric ancestors indirectly. For example, they study modern ape species and try to find if there is any measurement which corresponds with species longevity. In other words, they determine whether ape species with big brains, or big bodies, or big teeth, live longer than other ape species. As it turns out, body size correlates well with longevity in apes, so this has been used to concoct a story of our little, short-lived *Australopithecus* ancestors slowly becoming the larger and more long-lived *Homo* over the last five million years. Of course, this method depends on whether you believe that an apparent correspondence between body weight and longevity in living primates means that bigger hominids must *necessarily* always live longer than smaller hominids. Another problem with this indirect method of ageing our ancestors is that it becomes less useful as we get nearer the present – in the great scheme of things, the last quarter of a million years of human evolution seems so distanced from chimpanzees and gorillas that they may be poor points of comparison.

Another approach is to study the few hunter-gatherer societies still in existence. However, it is a big assumption – and an assumption which anthropologists sometime make – that modern hunter-gatherers are somehow representative of how all humans lived ten thousand years ago. After all, if they are so typical of the human race, then why did they not do what everyone else did, and start farming? Alternatively, a counter to this criticism is that most modern hunter-gatherers live in regions of the world to which agriculture has come relatively recently – perhaps two thousand years ago in southern Africa, or not at all in the Amazon basin, the Arctic and New Guinea. So maybe they are, after all, representative of the rest of humanity, but just happened to be living in a fortuitously unrepresentative location.

Even putting these concerns aside, modern hunter-gatherer societies do not tell a consistent story about survival into middle age. Put simply, in some tribes adults often survive to forty and well beyond, whereas in others they do not. Around the world there is tremendous variation in the longevity and causes of death in hunter-gatherers. Also, many do not seem to die of disease, starvation or predation (which are the causes of death most studied in animals), but violence at the hands of other people. Paralleling this variability in modern hunter-gatherers, most anthropologists have also long abandoned the idea that all pre-agricultural humans once existed in the same homogeneous 'state of nature'. Quite the opposite: many point out that one of the reasons humans have been so successful is that they have a tremendous flexibility to adapt their physiology and way of life to exploit new and varied environments. In this light, it would actually be surprising if survival to middle age was the same in all human communities.

The most reliable clues about how long those middle-aged pre-farming humans lived can be found in the human fossils that remain from that period, although estimating the age of long-dead people from the few fragmented fossils we pick from the dust is notoriously difficult. In fact, even estimating the age of recently dead adult humans is far from straightforward. Skeletal changes offer some clues – for example, changes in the structure of the seam between the two halves of the pelvis, or alterations in the microscopic architecture of bone. The pelvis method is accurate to within perhaps four years, but cannot be used beyond the age of forty, which is frustrating for anyone studying middle age. The bone-architecture method can be used in older humans, but may be inaccurate by as much as twelve years.

It is notable that forensic scientists find it relatively easy to determine the age of dead children because they can use their regular pattern of tooth loss and replacement as an accurate clock of their development. Yet, frustratingly for us, adults do not reliably lose or gain teeth. However, anthropologists have been able to exploit the fact that teeth wear down over the course of a lifetime to calculate ancient human longevity. First, they study teeth from the younger members of an ancient prehistoric population, who were still erupting new teeth. From these, it is possible to calculate how quickly their teeth wore away in the interval between the emergence of two successive teeth (and that interval does not vary much). If you then assume that this rate of wear also occurred in adults, it becomes possible to use the degree of wear in an adult fossil tooth to calculate the number of years between the eruption of that tooth and the death of the individual. Then it is a simple matter to work out the individual's approximate age at death.

These direct methods of measuring human age have shown that life was actually far from nasty, brutish and short for many ancient hunter-gatherers. In fact, the number of older adults in the human population had become impressive by the 'upper palaeolithic' era, which started 50,000 years ago. Patterns of human longevity can also be compared around the world, too – both *Homo sapiens* and *Homo neanderthalensis* had varying, and sometimes high, probabilities of reaching middle age depending on where they lived.

Yet the most striking result of direct measurements of the age-at-death of human fossils is the discovery that the number of older human adults actually *decreased* over the period when agriculture was adopted. This seems counterintuitive, especially when we are used to assuming that farming was a great leap forward in human development. So what happened?

The evidence we have seems to indicate that changes in diet are to blame. Agriculture necessarily reduces the variety in what people eat: they often have access to only one or two crops they can grow successfully, which limits their intake of vitamins and minerals and different kinds of protein. Of course, it also means that failure of those few staple crops can be catastrophic. Tying in with this, there is evidence from modern-day societies that farming actually reduces the amount of food gathered per amount of effort exerted. !Kung hunter-gatherers in southern Africa have been reported to work less hard yet be better provisioned than nearby agricultural communities, while Hadza hunter-gatherers are estimated to gather food for just five hours a day, in comparison with the day-long toil of nearby farmers. And tellingly, nearby agricultural communities often come to live with these tribes in order to survive the waves of famine which regularly sweep across Southern Africa.

Fossil evidence for the post-agricultural decline in diet is also accumulating. After the advent of farming, children's limbs became shorter and less robust, adults did not grow as tall and their teeth became smaller – and all of these changes occurred too fast to be explained by anything other than poor diet. Other more specific diet-related changes are apparent, too. The amount of abnormal growth of the outer coating of the teeth (enamel hypoplasia) increased, as did the incidence of porous skull bones resulting from iron deficiency anaemia (porotic hyperostosis). And bone mass, which is also influenced by diet, decreased over this time as well.

Another factor that reduced life expectancy in agricultural societies was infectious disease. When humans lived as dispersed, mobile hunter-gatherer tribes, there is no doubt that they suffered from some ancient diseases which have presumably plagued our species throughout our history – tuberculosis and intestinal parasites, for example. However, agriculture and settlement brought with them a terrifying range of new diseases. First of all, most early human permanent settlements were probably filthy and densely populated, and the lack of sewerage and the slowing of watercourses for irrigation would only have exacerbated the spread of illness. The most significant diseases in the world today are those caused by poor hygiene and population density, yet it is unlikely that these would have afflicted the old nomadic hunter-gatherer tribes on the open plains, who could simply walk away from their own filth.

The other factor in this increased incidence of disease was the new proximity of humans and animals. Before humans started to herd livestock, their main contact with animals was probably brief episodes of hunting, when the greatest risk was of being trampled or devoured. In contrast, livestock farming brought humans, animals and their various bodily excrescences into intimate, continual contact. Now, with farming, humans and animals are free to share bacteria, viruses and parasites as never before (think of avian and swine flu). Add to this the cocktail of previously buried pathogens exposed to the air by the plough and subsequently wafted into human lungs or absorbed into human wounds, and you can see that agriculture may have been far from the wonderful innovation it is usually assumed to be.

There is good evidence that, even today, patterns of disease change when humans switch from hunter-gathering to farming. For example, modern Turkana tribesmen have shown dramatic shifts in the bacterial infections from which they suffer when they have periodically made this transition in recent decades. And we have evidence of similar disease shifts in fossil humans, too. Some infectious diseases leave their mark on fossil bones – for example, Staphylococcal osteitis – and the fossil evidence suggests that these also became more prevalent with the coming of agriculture.

So why, if agriculture had such awful effects on human health, did it become the dominant mode of human life in such a short period of time? After farming, people worked harder for less return, ate a monotonous diet, lived in filth and died young, so what could possibly make all that suffering worthwhile? In fact, although agriculture was moving us into the modern period of human existence, the uncaring hand of natural selection was still acting upon us. For all its failings, farming does have some advantages. For a start, it requires much less land than hunter-gathering, so people can live in settled, small areas at far higher population densities – without agriculture there might still be only one million humans in the world. Also, when times are good in farming communities, the steady flow of calories means that people become very fertile and can produce many offspring. In short, farming makes human breeding more prolific, and in evolution that is what counts. It matters little that it makes people suffer and die prematurely.

So in many ways the last ten thousand years of human evolution may be seen as an aberration – a time when we strayed from a way of life we had practised for millions of years. If we want to understand how and why middle age evolved, it is obvious that the last ten thousand years of settled, farming life are little more than a distraction – a mere blink of the evolutionary eye, in which only limited alterations could have been made to the pre-existing pre-historic human life-plan. For 99.5 per cent of *Homo*'s tenure on this planet, we simply did not live the settled, farming life. And the evidence suggests that for much of that 99.5 per cent, middle-aged people were a large component of human societies. Crucially, this means that natural selection had many millennia to hone middle-aged humans into the wonderful end-product we see today.

4. Why is middle age so important? (A first attempt at an answer.)

It is time to make our first attempt at explaining what middle age is, and why we have it.

So far, we have been exploring the causes and origins of middle age. We have seen that middle age is part of a lifelong human developmental programme, a 'clock of life', in which thousands of genes interact to mould our bodies and brains into different forms at different ages. We have also considered how this programme has been sculpted by millions of years of natural selection, leaving us with our modern, human life-plan. We have seen that during middle age the processes of development come into conflict with processes of ageing and deterioration, and that this conflict gives middle age its distinctive flavour. And we have discovered how for most of human evolution – up until the advent of farming, at least – many adult humans survived into middle age and beyond. Because of this, middle age can now be considered to be a product of natural selection – something that helped us survive in the long unwritten past, rather than just a spurious, unnatural deterioration we only experience today because we now live 'too long'.

Yet to appreciate the value of all these findings, we must

reconcile them with what we see today – people's everyday experiences and perceptions of middle age. It is not a very scientific thing to say, but we need to find out what middle age is 'for' (or better, what advantages it confers on us). What are the roles for which middle-aged humans evolved? And do they explain why human mid-life *looks* so much like an ordered process? In other words, can we start to explain the triad of middle age I mentioned in my introduction: why are the changes of middle age so distinctive, abrupt, specific?

People are strange. Because of my background in zoology and veterinary medicine, I used to assume that human beings are just another species – intelligent, perhaps, but otherwise unremarkable. However, the more I have compared humans to other animals, the more bizarre and exceptional they seem. They stand on two legs, their life-stories are unique and contorted, their brains are bizarrely huge, their social structures are impenetrable, and their reproduction is weird to say the least. By any standards human are special, and the phenomenon of middle age is a crucial part of that specialness.

Even among primates – who are pretty unusual anyway – humans stand out. In some ways we have taken primate-ness to an extreme. For example, on the whole, primates survive well and live a long time, but we humans survive really well and live a really long time. We have also exaggerated other unusual primate characteristics, by undergoing puberty later than any other primate and giving birth to exceptionally immature babies who grow to adulthood with aching sluggishness.

In other ways, though, we have entirely broken the primate mould. Human females produce babies more often than other female great apes do, and they suckle each baby for less time.

As a result, although a great deal of resources are dedicated to each child, human females often care for several offspring at the same time – unlike our close primate relatives, women do not wait until one baby is grown and independent before they conceive another. Another clear example of human unusualness is that human females often survive far beyond the age at which their fertility ceases. Men also live beyond that age, remain technically fertile, yet often effectively 'self-sterilized' by staying with their now-infertile partner. Men then confuse things even more by dying younger than women, despite retaining the potential to produce offspring long into old age. And of course humans of both sexes frequently take the evolutionarily inexplicable decision not to produce offspring at all. Chimpanzees do none of this.

Frustratingly, if we try to study the origins of the unusual human life-plan, there is very little direct evidence to guide us. Our closest living relatives – chimpanzees and gorillas – are just too different in their reproductive biology and social organization. And the other point of comparison for us post-agricultural humans – modern hunter-gatherers – are just too similar. Around the world the unfolding of the human life story follows a fairly standard format, even with respect to the things anthropologists love to study, such as division of labour between the sexes, systems of incest-avoidance, social recognition of marriage and mechanisms of subjugation of women.

In fact, clues as to how the human life-plan evolved can be found if we look in other directions. Evolutionary biologists are now starting to focus on our more distant primate relatives, and to create mathematical simulations of human populations. And suddenly, humans are beginning to make more sense. As it turns out, there are two really striking novelties about the human life-plan: adolescence and middle age. It is sad that these two most astounding human innovations – the teenage years and middle age – are so often viewed negatively, or at best dismissed as mere transitional phases between other, 'more important' stages of life. Yet between them these characteristically human stages encompass perhaps half of our entire lifespan. Does it really seem likely that half our lives should be spent in problematic, negative, transition?

Of course, adolescence and middle age are linked by the simple arithmetic of the years. Most teenagers have parents who are middle-aged and, as it turns out, it is no coincidence that people in these two phases of life usually live in direct contact with each other. I have argued elsewhere that humans evolved the teenage years – a uniquely human extra ten years of development – so that they could perfect the enormous, all-conquering brain which has been the secret of our species' success. There is a great deal of evidence for this: in fossils, on the psychiatrist's couch and inside the brain scanner. But there is another side to this story, and that complementary side is middle age.

The extension of human development into the second decade can be seen as an exaggeration of a tendency already present in human children. Young human children require far more resources and care than other infant apes, and the reason for this is that their brains are burning energy and consuming new ideas at a ferocious rate. Because of the demands placed on us by our brains, human life runs at a different pace from that of other animals. Our species is a high-investment, information-intensive economy – everything revolves around driving the growth, maturation and productivity of that huge, demanding brain. As a result, a human life is an exercise in investment – adults plough enormous resources into children's developing brains because those brains are just so darned useful in later life.

Biologists call this 'parental investment', and humans are parental investors de luxe. By any measure, parental investment is more protracted, complex and demanding in humans than in any other animal. Of course, this explains the *cri de cœur* of some exhausted middle-aged people who feel as though parenthood will never end. Yet we should also realize that without those demanding children and teenagers of ours, we would probably never live to see forty. It is now believed that because human offspring grow so slowly, natural selection has dictated that there is a time for us to stop breeding and instead concentrate on the offspring we already have. And this allotted time to stop breeding and start caring usually arrives during middle age.

Many studies suggest that parental investment is the most important single factor in the reproductive success of an individual human – their ability to produce their own mature, successful offspring. Indeed, we have now evolved to a point where parental investment is even more important to humans than fertility (which may explain why trying to get pregnant is such a ridiculously unpredictable exercise in our species). And some time in middle age, the supreme importance of parental investment in our existing children becomes so great that we stop making any more. Paradoxically, breeding itself becomes a harmful distraction from the important job of being a mother or father.

If parental investment has driven the evolution of middle age in humans, then we must find out what it actually entails – what it is, precisely, that adult humans provide for the next generation and their rapidly swelling brains.

We give our children two main things – the first of these is food. It takes a staggering amount of calories, protein and other nutrients to manufacture an eighteen-year-old. In humans the need to gather food to feed large-brained children is an overwhelming pressure. For example, eighty-seven per cent of the energy used by a resting newborn human baby is being burnt in its developing brain – a demand which other species' parents simply do not have to satisfy. And remember, ten thousand years ago somebody had to hunt or gather all of that energy. Even once children can walk and run, we do not let them do much food gathering. This may make them seem like a burden on human societies, but it also protects them from the dangers involved in acquiring food. Compared to other primates, childhood mortality is low in humans, and this is partly because our children are protected from the need to hunt and gather. The other reason why relatively few human children die is that when they fall temporarily ill, food is still provided by others, whereas in many species a 'temporary' illness often leads to death by starvation.

In fact, the entire system of food acquisition in humans is unusual. Most primates eat large amounts of food which is low in nutritional quality yet readily available – wild chimpanzees simply pick up most of their food. In contrast, humans had to change their eating habits, probably when climate change caused the expansion of the African grasslands approximately two million years ago. In response to the drying of Africa, humans stayed out in the open (in a most un-primate-like way) and foraged across large geographical ranges to find uncommon, inaccessible, high-value foodstuffs. Among primates, humans are unusual in specializing in foods which are buried, shelled, poison-coated or which run faster than we do. Humans focus on varied, valuable foods which require skill to acquire (finding, digging, peeling, outsmarting) and this is probably one of the main reasons we became so clever.

Of course, learning these skills takes time. By the age of five, chimpanzees can already gather enough food to feed themselves and very soon they are moderately efficient foragers, and will remain so throughout adult life. In contrast, humans probably make no net input into their community's food budget until they are at least twenty years old, but then something fascinating happens. In modern hunter-gatherer societies, the ability of individual adults to gather calories just goes up and up as they learn the skills of hunting and gathering. Studies suggest that between the ages of twenty and thirty-five the rate of acquisition of calories by hunter-gatherers quadruples. Eventually each human can acquire calorific foods far faster than any chimpanzee – in other words, they take a long time to learn their craft but eventually they get extremely good at it. And from the point of view of our story, the most gratifying piece of information is that the ability of humans to gather food peaks at the age of forty-five (by which age most chimpanzees are dead). At forty-five, hunter-gatherers will be losing strength, bone mass and agility, but their years of practice and experience mean that they still outshine their younger peers. In terms of gathering resources for their community, middle-aged people have always been the best.

Provision of food for offspring has had some remarkable effects on human society. The process is not a simple one in which all adults contribute equally to feeding the young; instead, some individuals hunt and gather more than others. Most primate communities can amble around en masse simply picking up food, but in humans this is not possible.

Advanced hunter-gathering requires skill and mobility, and bringing along the kids and both parents simply does not make sense. So humans had to specialize – some of the community went out to find food while the rest stayed behind. In addition, humans operate a system whereby individuals other than parents contribute resources to growing children. Modern hunter-gatherer parents receive help from other adults – a study in South American hunter-gatherers suggested that each couple receives help from an average of 1.4 other adults when providing for their children. In fact, it seems that the general trend across the human race is for adult men, and especially middle-aged adult men, to be the main providers of additional calories to parents. And presumably it is also these middle-aged men who 'train' the younger men who will eventually take on their role. (Of course, this implies that middle-aged women are not as significant a source of food provision. We will return to what *they* might have to offer later.)

So human patterns of food gathering are exceptional, with middle-aged men in the key role. Continual redistribution of food appears to be central to human societies, and may indeed be the basis of the intense sociability of our species, which in turn is probably another important reason why our brains are so big. In this way, an explanation for how distinctive elements of human biology evolved in tandem begins to emerge. Our enormous brains require so much energy when we are young; that energy must often be provided by non-parents; this provision binds us into a tightly social arrangement; and that sociability requires yet more brain capacity. We humans are locked in a virtuous circle of intelligence, skill and sociability with middle-aged people as its driving force.

*

Creating a functioning human child requires more than just food, of course: they also need information. As we have seen, much of this information is contained in our genes. In many animals, *all* the information is in the genes, but in complex animals such as mammals, and especially humans, young individuals receive an important tranche of information in another form – they learn it from their elders.

This non-genetic inheritance of learnt information is extremely important because it includes a great deal of information we find useful in our everyday lives – in other words, it helps people survive and breed. A wide array of knowledge, skills, values, attitudes and goals is passed down the generations – you might call this assembled body of information 'culture', in fact. And because humans collect food by sophisticated means, care for their offspring in complex ways, and have intricate social interactions, there is a lot of culture for young humans to learn. We are born with little innate knowledge of how to do, well, *anything*, so intergenerational transfer of information is essential. Children who grow up starved of contact with other people simply do not acquire the culture of human existence and find it difficult to function normally in later life.

At first sight, passing down information in this non-genetic way seems precarious. Unlike the certainty inherent in the bequeathal of the DNA code to offspring, the perpetuation of a culture requires an alarmingly ad hoc mixture of verbal advice and setting an example to the next generation. If culture-transmission were to fail for even a single generation – because of an environmental or social calamity, for example – subsequent generations would be irreparably compromised in their ability to survive and prosper. Genes can last for millions of years, but thoughts slip away from us so easily. However, the evanescent nature of the intergenerational transfer of information is also its greatest strength. New skills or insights gathered in a single moment by a single individual may be rapidly perpetuated throughout that individual's kin, descendents, allies and friends.

In fact, although culture changes over time, it seems surprisingly resilient. Human societies are so driven to pass on their beliefs and practices to the next generation that there are very few examples of their failing to do so. Of course, being the only species with bona fide language is a great help. It means we do not even have to show our young folk what they should do; we can just tell them. Humans are unique in being able to verbally articulate anything that comes into their heads, which is why human culture has risen so far above the simple cultures of food-gathering, tool-use and vocal communication seen in other intelligent species. This is why only in humans can adult females criticize their daughters for wearing too little on a winter's night, and adult males bore their sons with the minutiae of seventies' rock music.

And who is largely responsible for this cultural transition if not the middle-aged? Human life-plan theories now suggest that this information-conveying role is a major reason why people survive beyond their reproductive years. We have already seen that it is middle-aged humans who possess the appropriate combination of experience and physical vitality to be the species' top providers. Well, now we can see that even when that physical vitality starts to wane – when muscle shrivels and bone thins – the older members of human communities will still be the most experienced, too. The world is full of middle-aged people teaching and training young adults, even though those young adults are often brighter and quicker than they are. This gives middle-aged individuals a tremendous inherent value.

If Darwin were still with us, a sage smile would rustle his beard and he would tell us that this means that natural selection can act upon the middle-aged. Because the middle-aged stage of human life promotes our offspring's success, it evolves, even though it falls beyond the age at which most people stop making babies. Cultural transmission gives people an evolutionary importance far beyond that conferred by their crude ability to breed

So what evidence is there for this theory of middle-aged cultural transmission? Well, anecdotally, we all know that as people get older they develop a liking for giving advice and making their opinions known. And as they creep into old age, this desire for conveying experience can become something of a compulsion, and occasionally an irritating one at that. At some point in middle age the dual realizations dawn on us that not only do our experience and knowledge constitute an increasing proportion of what we can contribute to society, but also the amount of time we have left to convey that experience and knowledge is shrinking menacingly. In middle age there still seems enough time to convey measured and balanced perspectives to younger generations, but in old age the process can become a desperate, frustrating struggle to heap information onto a younger generation which no longer seems to be listening. So much to say: so little time.

Of course, those sorts of observation are entirely unscientific, but the idea of middle-aged human information repositories chimes well with the unusual structure of the human life-plan. Humans do not often die during middle age, even though we breed far less frequently and obvious signs of bodily deterioration have appeared. As we will see in later chapters, our brains, and especially our memory and language, undergo little deterioration in the fifth and sixth decades of life, and of course this is exactly what one would expect of beings whose main function is to store and impart information – be they parents, other relatives, friends, or workplace mentors. In this context, it is notable that studies suggest that lack of interest in supporting younger people can be an early predictor of illness and death in the elderly. It is as if once we cease to transmit our thoughts to the young, natural selection finally loses interest in us.

And now some neuroscientists believe they have found the very parts of the human brain involved in this compulsion to cultural continuation. In particular they suggest that two freshly evolved, interacting regions of the human brain may be involved – 'region 10' of the frontal lobes and the spindle cells of the anterior cingulate gyrus. These two regions are thought to be central to the process we all undergo when we feel we have failed at something – instead of giving up or asking for advice, the initial reaction of most adult humans is to analyse what we did wrong and work out how to be more successful next time. It is claimed that this brain circuit is the basis of middle-age introspection (which we will look at in more detail later), and some believe that it also drives our compulsion to transmit knowledge and experience to the young – to allow them to learn from our mistakes.

We are at a turning point in our exploration of human middle age, as we now know not just how but also *why* we evolved middle-aged people. We have seen that their superior provisioning skills and experience are essential for the survival and continuation of the species, and that without middle-aged people the high-maintenance human way of life simply would not work. The reason the changes of middle age are so distinctive, abrupt and specific is that they are not the symptoms of gradual uncontrolled deterioration. Instead, they were built into us by natural selection, and we gain huge physical and cultural advantages from them.

We are now free to go on and study the nitty-gritty of modern middle age – what changes in us once we have hit forty, and why it changes. But before we do, I should say one more thing about that vast brain which has been such a distinctive feature of our evolution. Not only has our brain allowed us to develop skills and cultures of tremendous complexity, but it has also given us self-awareness. Studying the ancient history of middle age has a singularly spooky aspect to it. All those middle-aged people over the last two million years were not unthinking automata, acted on by the callous hand of evolution. It is important to remember that they were probably just as clever and self-aware as we are. For two million years middle-aged people have thought the same thoughts that middle-aged people think today. Most have probably wondered about what their role in society would be once the wrinkles started to appear. Many will have pondered what they should do with the rest of their life. I am sure that almost all will have fretted about what they felt they had missed out on when they were young.

Our perspectives on middle age have now changed for ever. No more will we think of prehistoric peoples as a disorganized rabble of (mostly young) hairy folk, stumbling their desperate, inchoate way to their next meal. Now we can imagine them whipped into a lean foraging and child-raising machine by an *über*-caste of the pushy, self-confident and occasionally even self-righteous middle-aged.

5. Saggy? Wrinkly? Grey? Why?

Keep young and beautiful,
  It's your duty to be beautiful,
Keep young and beautiful,
  If you want to be loved.

Al Dubin, lyrics for 'Keep Young and Beautiful', 1933

Now for the bad news.

As you may have noticed, I am upbeat about middle age – keen to sing the praises of a unique and essential human innovation which has evolved to be productive, positive and potentially enjoyable. However, I cannot deny that there are negative aspects to middle age, but I hope to compress most of them into this one, single chapter.

Around the age of forty it seems as if our bodies have a change of heart. Built into our developmental programmes are some relatively sudden and distinctive changes which do not so much creep up on us as grab us by the shoulders and forcibly point us in a new direction. Of course, we do not suddenly 'get old' at forty – whatever the jokes inside birthday cards say. The nature of human middle age has never been about getting old. However, the change of direction can be remarkably abrupt nonetheless, and the rapidity of this change shows that it is an organized, directed process. Within a few years, the emphasis of our lives alters for ever.

The forces of evolution, and their hand in the creation of middle age, have ensured that it affects different parts of the body in different ways. Although middle-aged humans are destined to play important roles in resource provision and information transfer to the young, other things we do become less important, and one of these is looking good. In evolutionary terms, it is the production of babies by human heterosexual couples which counts (even if you are homosexual, single, or have chosen not to have children, you still almost certainly inherited all your genes from a child-bearing heterosexual couple). And as we get older, our potential for producing children gradually decreases, simply because we each have fewer years left. Because of this, while suitors may still find us attractive, they are often not seeking us out as potential co-parents, and thus are no longer choosing us because of our youthful, fertile-looking appearance. Natural selection therefore loses interest in some aspects of our appearance. As long as we are still recognizable as ourselves, and are not so decayed or repulsive that any potential or pre-existing mate is driven away, our appearance may be allowed to deteriorate. After forty, the superficial aspects of you and me can to some extent go hang, both figuratively and literally.

Most superficial of all the body tissues, it is the skin which suffers most in middle age. This is the bit of 'soma' which seems most 'disposable'. Within an alarmingly few years the skin becomes noticeably less, well, *beautiful*. These changes take decades to be completed, but their onset can be sudden and depressing. Many of us still want to look like Luke or Leia at a time when nature does not seem to care if we end up looking like Yoda. Of all the elements of middle age, it is the changes to our skin which seem most cruelly unforgiving. So what can we do to fight this cutaneous collapse?

The skin is a large and complex organ. It consists of two main layers: a visible surface layer called the epidermis, and a thick, resilient, elastic deeper layer called the dermis – an arrangement not unlike an attractive carpet laid on springy underfelt. The epidermis is continually self-replenishing, its deepest part always producing new cells which gradually become tough and waterproof as they migrate upwards towards the surface over a period of approximately forty days, and thence are shed, dead, into the air. So the top layer of the epidermis, that luminous veneer we love to touch and kiss is, I am afraid, a moribund waste product. The dermis, in contrast, is a more vivacious, fleshy layer with blood vessels, nerves and the cells which make the fibres which give skin its strength. The epidermis and dermis join forces to keep our insides in and marauding outside-type-things out, and in places they also form other structures. For example, tiny tufts of epidermis may burrow down into the dermis, while still retaining their connection with the outside world. In doing so they create those two unique features of mammalian skin – follicles which extrude hairs, and glands which secrete sweat and sebum. The skin is an amazing organ, performing many important functions in elegant ways, but as far as middle age is concerned the news about almost every aspect of the skin is, to put it bluntly, bad.

In early middle age, the skin starts to lose its elasticity. This failure of the skin to bounce back may be unpleasant but in fact it has an extremely long evolutionary story behind it. One of the most important junctures in the history of life on Earth came when individual cells clumped together to make multi-cellular organisms. For most of Earth's history, organisms were microscopic, single-celled flecks with little scope for becoming complex. Only relatively recently did cells gang together to make animals, plants and fungi, and when they did, they needed stringy, fibrous molecules to hold them together. The stringy molecule which holds many animals together is a protein called collagen, and along with its springy colleague elastin, it makes up an appreciable proportion of animals' body weight.

The dermis is full of collagen and elastin fibres and it is these which make it so tough and elastic. And in middle age deterioration of collagen and elastin in the dermis causes many of our cosmetic worries. The number of cells which make these fibres declines, so the rate at which they are produced and replenished decreases. Also, the fibres may be actively broken down, or accumulate in disordered, abnormal tangles, making the skin weak and unattractively rigid – rigidity is not a good thing because it leads to wrinkles. In addition, the blood supply to the dermis decreases in middle age, while the epidermis thins and its waterproof cells stick together less well, resulting in pale, translucent skin which is a poorer protective barrier.

Although some degeneration of skin collagen and elastin is unavoidable, there are certain things which make it much worse. Perhaps the most important of these is sun exposure, and there is even a term for this phenomenon: 'photoaging'. Ultraviolet light has been shown to stop collagen fibres connecting normally. Also, it increases the production of enzymes which destroy elastin – and experimental administration of chemicals which suppress these sunlight-activated enzymes has been shown to reduce wrinkle formation. However, smoking may be even more destructive to human skin than sunlight – it reduces blood flow to the skin, the smoke itself dries the facial epidermis, and wrinkles form where the mouth puckers to hold a cigarette and where the eyelids squint to protect the eyes from smoke. All in all, considering how destructive these two factors are, it is surprising how we instinctively equate tanned skin with health, and sometimes even think that smoking looks 'cool'.

There are yet more factors which hasten failure of collagen and elastin. Gravity is an unforgiving culprit, but there is not a great deal we can do about that, given that we spend most of our time upright. Sleeping position, however, is something we can change. The imprint of bed sheets on our recently roused face may be funny when we are young, but all that facial creasing gradually becomes more permanent in later life. This is especially important because it is facial skin rigidity which most effectively makes us *look* old – the inelasticity which makes our eyelids droop, or our upper lip turn in and our lower lip protrude. If you want to look young, sleep on your back, and for similar reasons, cut down on extreme facial expressions. Finally, stress can also affect our appearance, as it leads to the release of glucocorticoid hormones which are known to cause the breakdown of collagen in the skin.

As you can see, there are many things which hasten wrinkliness, and we believe that external influences and our genes contribute approximately equally to our tendency to lose skin elasticity. Obviously, this means that we can reduce skin ageing by avoiding sunlight, cigarettes, stress and the pillow nemesis, but is there anything we can do once rigidity has set in? Unfortunately, there is no miracle cure for an ageing collagen-elastin system, although of course that has not stopped an entire mid-life cosmetic industry being established around such 'cures' – an industry which frequently makes claims which verge on the fraudulent.

Vitamin A is one potential treatment with some limited

effect. It may encourage skin collagen synthesis, but is probably more likely to work when taken orally than when slapped on in ever-more-expensive skin creams. Paradoxically, it can have the effect of making the superficial layers of skin look dry, so it must be combined with a moisturizer to stop the punters demanding their money back. Using vitamins C and E to reduce skin rigidity also makes sense – C does play a role in collagen production, and E neutralizes destructive reactive oxygen species – but neither has been demonstrated to have any effect on the elasticity or appearance of middle-aged human skin. The skincare products I am most doubtful about are the creams which contain collagen, and occasionally elastin. One of the skin's main *raisons d'être* is that it stops chemicals entering the body, and chunks of processed collagen or elastin are unlikely to penetrate far through the epidermis, and certainly not far enough to enter the dermis where they are 'needed'. Even then, I am not sure I would want that stuff magically incorporating itself into my dermis – especially as it is probably best not to think too much about where the cosmetics companies actually get all that collagen.

So wrinkly skin can be partially avoided by changing your behaviour rather than your face cream, but once you have it there is little you can do except surgically stretch it flat or use botox to paralyse it into a smooth, expressionless mask. Instead of wasting your money, give your children some good advice about the future of their skin. I know from long, painful experience that many children are extremely unwilling to eat healthily, stop pulling faces, and keep out of the sun, and they are also unlikely to be too worried about the risks of looking like a brown leather handbag in their inconceivably distant middle age. However, where skin is concerned, a virtuous youth is far more effective than a mis-guidedly profligate middle age.

Loss of skin strength and elasticity affects each of us differ-ently. It can make earlobes sag and noses lengthen. And in women, it is a major contributor to the southward migration of the breasts. Unfortunately, mammary glands are not designed for long-term structural integrity. Unlike any other major organ, they are neither protected by bone nor sup-ported by sheets of muscle. They are entirely derived from the skin (they are probably modified apocrine sweat glands), so their only physical support is a slight elaboration of the collagen and elastin present in any other piece of skin. This situation is not helped by the fact that the main components of breast tissue, fat and milk glands, are themselves not struc-turally robust. Thus, the loose fibrous capsule of the breast gradually deteriorates and loses its fight against gravity – and it is for this reason that small-breasted women can feel a certain satisfaction as their more curvaceous peers sag at a younger age. Previous pregnancies and breastfeeding also hasten the decline because they will have caused the breasts to swell and recede, which stretches the fibrous capsule further. The other major 'unsupported' human organ is the testicle, and many middle-aged men notice their testicles creeping into the distance as the scrotal skin stretches. Unlike breasts, testicles do at least have a strand of muscle attaching them to the torso, but even this weakens with time.

The changes taking place in middle-aged skin are not restricted to loss of strength and elasticity, however. Other malign forces are at work below the dermis to adversely affect our appearance. Many people suffer shrinkage of their sub-cutaneous fat pads, especially in the face, which previously gave their facial contours a full and smooth appearance.

Frustratingly, this occurs at a time when many of us are trying to control our weight, and rapid weight loss during middle age can hasten fat pad loss and end up making us look older, and even cadaverous.

Another part of the body which often betrays our age is our hands – and so far cosmetic surgeons have found few solutions to this problem. The reason for this is that although there is some subcutaneous fat in our hands, much of the padding beneath the skin is provided by the muscles which perform the hand's complex movements. In middle age, those muscles shrink and may be partially replaced by fibrous tissue, giving the hands a wrinkled, skeletal appearance – an effect emphasized as the fingernails become thinner, brittle and ridged.

In middle age, skin secretions decline too, including watery 'eccrine' sweat, with the result that our ability to tolerate heat declines. Also, secretion of oily sebum decreases, and this can adversely affect the delicate balance of moisture in the skin. This change is much more dramatic in women, and exposure to a moisturizer as crude as tap water has been shown to rapidly increase skin pliability in middle-aged women, but not men. Of course, this suggests that it is more beneficial for women to use moisturizer, especially late at night before they wedge their face against that disfiguring pillow. In fact, moisturizing is probably one of the few cosmetic interventions which actually works, although it is only a short-term measure which must be performed frequently because it affects only the most superficial layers of the epidermis – layers destined to be discarded and replaced within a few days.

The final chapter of the sad story of middle-aged skin relates to skin pigmentation. The colour of human skin results from the presence of pigments – melanin and carotenoids – backed up by the warm glow of red haemoglobin in the blood vessels of the dermis. I have already mentioned that dermal blood supply declines in middle age, and the skin can also be marred by the formation of fragile, tortuous blood vessels. However, the most dramatic change in skin coloration is that the distribution of melanin pigment becomes uneven and irregular. The total number of melanin-secreting cells decreases during middle age, and this conspires with reduced blood flow to make most of the skin look paler. Despite this general pallor, focal pigmented spots often appear. These may be called 'liver spots' or 'solar lentigos' and while they have nothing to do with the liver, their second name is more accurate. The main cause of these spots is sun exposure – the skin's natural tendency to deposit protective melanin is permanently stamped into patches of our ageing skin. So once again, we pay for our youthful frolics in the sunshine when we reach middle age.

The most remarkable elaboration of mammalian skin is hair – produced when tiny bulbs of epidermis start to extrude long, fine, pigmented columns of the protein keratin.

Or at least, those columns are pigmented when we are young. Hair greying, which often starts tentatively in our thirties, continues apace in middle age and most of us are predominantly grey by the age of sixty. Of course, people do vary in the age and rate at which they go grey, and we believe that hair greying is controlled almost entirely by our genes – so unlike skin ageing, there is little you can do (beyond dyeing it) to hold back the grey tide. Even within our species there are consistent differences – Caucasians go grey earlier than other races, for example. However, it is important to appreciate that, unlike many features of human middle age, hair greying is not a distinctive feature of humans. One only has to think of the silvery muzzle of an elderly black labrador to realize that many animals go grey as they get older. However, the great apes are especially enthusiastic greyers, and the dramatic silver back of old male gorillas or the grey margin around the faces of wizened old chimpanzees does make me wonder if we primates might use grey hair as a specific signal of age. Maybe we even use it to signify our importance as repositories of cultural information.

The greying of human hair is not a homogenous process. It often starts at the temples, or in small patches along the fringe, and then spreads across the head in a piecemeal fashion. Many people have extensive regions in which grey hairs are intermixed with pigmented hairs, often for decades. Body hair usually greys later than head hair, and male facial hair can show a very complex pattern of greying, perhaps echoing the role of beards and facial hair as signals of maleness and dominance in many other primates. And finally comes one aspect of middle age that no one likes to think about. Yes: pubic hair goes grey, too, but thankfully later and sometimes not at all.

Scientists are gradually moving closer to understanding the processes which underlie hair greying, although this has not yet provided any inspiration as to how to delay it. Unlike skin, the colour of hair is entirely due to melanin pigments – even red hair gets its colour from 'phaeomelanin' – and it is the loss of melanin in individual hair follicles which makes hair appear grey. Most melanin in the body is produced by unusual cells called melanocytes, which form near the developing embryonic nervous system and then throng throughout the length and breadth of the body. One of the places they congregate is the hair follicle, where they infuse melanin into the growing keratin hair shaft.

Hair follicles do not secrete the same hair for ever. Every five years or so hairs are shed, a new one starts to grow, and new melanocytes bustle into the follicle to produce pigment for it. In middle age melanocytes are not recruited into the follicle as efficiently as before, and once there they survive less well. Rather than being a global, whole-head phenomenon, this depletion mysteriously affects individual follicles, which is why most of us go through a 'salt-and-pepper' phase when non-pigmented follicles are scattered among pigmented ones. Although scattered, melanocyte failure is irreversible, so plucking a grey hair will not elicit the sprouting of a pigmented one (nor will it provoke a cascade of many grey hairs from the single follicle, as the old wives' tale would suggest). The failure of follicle melanocytes could be a pre-programmed change of middle age, or it could result instead from follicular 'exhaustion' caused by the extreme metabolic demands of making hair for all those years. Making melanin generates reactive oxygen species as a by-product, and we have already seen that these are implicated in the ageing process. Hair follicles can also accumulate high concentrations of hydrogen peroxide, and rather than directly bleaching the hair in a platinum-blonde sort of way, this too can be toxic to melanocytes.

As well as the colour of hair, its distribution also changes in middle age. I am not going to discuss male pattern baldness, because that often starts as early as the twenties and so is not really a characteristic feature of middle age. However, striking alterations do take place in hair texture and distribution over much of the body. Body hair comes in two varieties: fine, downy, almost invisible 'vellus' hair, and coarse, often pigmented 'terminal' hair. Body hair follicles are induced to change from wafting out vellus hairs to jabbing forth terminal hairs by androgens – male sex hormones such as testosterone. This is why men and women differ in their body and facial hair – in fact, their hair distribution is almost identical, but in women most of it is vellus whereas in men much of it is terminal. Also, different hair follicles have different sensitivities to androgens, which explains why most women grow visible hairs in their armpits but not on their chins.

Male body hair is the clearest evidence I can think of that our developmental programme continues throughout our lives. Men do not establish a final 'definitive' distribution of body hair at puberty, by eighteen, or even later. Instead, their pattern of hairiness continues to develop well into their third and fourth decades, and often beyond. The 'clock of life' is very obviously still ticking. Until middle age this mainly manifests as a gradually thickening and expanding forest of body hair, but in middle age the continuing maturation of male body hair takes some unattractive detours. Everyone has hairs growing in their ears and nose, but usually they are invisible vellus hairs, presumably with a minor protective function. But in middle-aged men they often become coarser and longer, as do the eyebrows, and all those strange grooming implements one sees in newspaper advertisements suddenly seem like a good idea. Women, too, develop more terminal hairs, but far fewer than men. However, they often consider the occasional terminal hair on their upper lip or chin more of a cosmetic insult than a man would – he has thousands already. One irritating paradox about middle-aged body hair is that many of its least-wanted aspects are caused by circulating male sex hormones: yet this is often a time when these hormones are actually in decline, and as we will see later, some men wonder if this explains their loss of libido. Middle-aged hair is a contrary thing.

*

It is all very well to dissect the way in which natural selection loses interest in the appearance of humans in their fifth and sixth decades, but what actually worries every one of us is how it makes us feel about ourselves, and how other people react to us.

There is little doubt that people are pre-programmed to assess the age and beauty of others, and that we use many of the phenomena described in this chapter as our cues. For example, studies of the visual system show that faces with a youthfully homogenous skin tone are more likely to attract our gaze, and that our visual 'dwell time' on these faces is longer. Also, questionnaire-based studies show that skin homogeneity is taken as a sign of youth and attractiveness, whereas wrinkles, grey hair and reduced lip size lead to perceptions of increased age. In fact, we seem subconsciously to dig even deeper than that – using skin pigment and texture as independent indicators of health and age, respectively. Studies comparing identical twins show that those who smoke, or have received greater exposure to ultraviolet light, are scored as being less attractive. Other twin studies have demonstrated that increasing levels of body fat in people younger than forty make them look older (to people asked to guess their age), whereas increases in body fat in people over forty make them appear younger. People's assessment of our skin is surprisingly consistent.

Most of these studies have examined the appearance of women – in other words, asking men and women to assess the youthfulness of female faces. I suspect this is not because of any sexism inherent in the research community, but because women's worries about ageing skin and hair appear more pressing. No one likes their skin or hair to age, but it does seem to worry women much more than men. Many people comment on the 'unfair' asymmetry in attractiveness between the sexes, especially as we get older. Studies show that men's facial wrinkles and grey hair are viewed less negatively by both sexes than those of women. In part, this is fortunate because men's facial muscles are more mobile and thus their faces wrinkle more quickly in middle age. However, I do not think the discrepancy in our perceptions of ageing in the two sexes can be explained solely by an innate acceptance that men get slightly craggier than women.

Instead, what we find attractive in someone depends very much on whether they are male or female. We will return to this essential inequality in human ageing later, but suffice it to say that I believe it explains the apparently unfair way we view middle age in women and men. In the modern, politically correct, science-oriented world it is easy to forget that some things in life are simply unfair, and age and beauty are two such things. Some middle-aged people look young and some look beautiful. Some look both and many look neither, and we simply have to accept it. At least middle-aged people in the developed world look younger now than they used to – hair dye, good diet, quitting cigarettes and working indoors have worked wonders for us.

However much we may not want to admit it, perceptions of superficial attractiveness are hard-wired into our brains. But what happens between partners as they grow older – what holds a human couple together? Does each partner slowly develop a sexual and romantic taste for older men or women as they enter middle age, which conveniently matches the age of their partner? Or do they somehow superimpose an outdated, youthful vision of their partner on top of what they actually see in front of them? Or, is human monogamy no more that a social invention, and would middle-aged people really rather ditch their ageing partners and find someone younger? We will think more about these challenging ideas later on.

Changes in our superficial appearance also profoundly affect our self-image in middle age, and this is largely due to an accident of the chronology of the human life-plan. When we are children and teenagers we 'feel' young, and one of the main reasons we feel young is that we believe we 'look' young, too. We have smooth, homogenous skin, and thick, pigmented hair, and this stands in stark contrast to our parents who are usually, by the time we are getting old enough to notice, showing signs of middle-age changes in their skin and hair. Even once we reach our twenties and thirties, I believe that we still class ourselves as 'the young' because we remain smooth and richly-coiffed, even if we ourselves are now more burdened by the pressures of life. However, once we pass forty we start to notice changes in our appearance which we have always instinctively associated with (horrors!) our parents. We humans are a very self-aware species, and creeping changes in our appearance as middle age approaches are a crucial factor in changing our perceptions of who we are.

So that was the bad news, largely crammed into one chapter. Our conclusion: life is not fair, especially where skin is involved. Changes in superficial appearance during middle age seem a simple, straightforward part of life, but now they have opened a sexual hornet's nest to which we must shortly return. However, before that, we will examine another aspect of our middle-aged appearance, and you may be surprised to find that I am more positive about that one: fat.

6. Middle-aged spread is normal, isn't it?

Something weird is happening to the weight of our species. As we are told almost every day, increasing levels of obesity are a dangerous problem in the developed world. People are fatter than ever before and this is making them so ill that we could be about to witness the first decline in life expectancy for centuries.

Yet we should not be surprised by this phenomenon. Obesity is simply the inescapable result of an imbalance between calories consumed and calories expended. There is no more to it than that. Many people lead more sedentary lives than their predecessors did several decades ago, but I suspect that the major culprit in modern obesity is food density and availability. Many humans now have ready access to a large quantity of high-fat, high-sugar food, and we enjoy eating it. We evolved in an environment free of Jaffa Cakes, so it should come as no surprise that we are unprepared to cope with their temptations. We are even managing to make our children fat and that is a real achievement – it is actually quite difficult to make immature mammals fat because they expend so much energy growing and playing.

However, we should not overstate the importance of the

'new obesity' in our investigation of human middle age, nor let it obscure our real interest – how and why human body shape changes in our fifth and sixth decades. Mid-life weight gain is a distinct, different phenomenon: it was evident long before childhood obesity became commonplace, and it causes a separate set of problems. Also, it is often viewed as normal, natural, inevitable and acceptable. So is it?

Fat is impressive stuff. Adult humans often carry between ten and fifteen kilograms of adipose tissue and this amount holds enough calories to keep us going for two to three months. It is hard to think of a more efficient way to protect yourself against future scarcity.

In the developed world, eighty per cent of people put on weight in their fifth decade. One study suggests that over the fifth and sixth decades of life, percentage body fat increases from 23.6 to 29.3 per cent in men and from 33.4 to 37.8 per cent in women. The average weight gain is one gram each day, and although this may not sound like much, it soon adds up when it accumulates over the years. That one gram also demonstrates how easy it is for tiny but continual miscalculations in our food/exercise balance to have enormous effects.

But these are average values, and they hide complex patterns. For example, socio-economic and educational statuses have complicated effects on middle-aged fatness, too. Women 'higher up the scale' are leaner in early adulthood and remain so, although their fatness peaks earlier in middle age. Conversely higher-status men do not show the same tendency to slimness in their youth but their fatness peaks later than lower-status men's. In developed countries, men's weight peaks around the age of fifty-five and women's at sixty-five. Also, during the current 'obesity epidemic' it is noticeable that lean people have not got fatter. They are just as lean as ever, but average fatness has been skewed upwards by fat people becoming disproportionately fatter. The trends are complex in developing countries, too. Some show an opposite pattern to developed countries, with fatness decreasing between the ages of twenty and fifty, whereas in other countries middle-age obesity does occur, but it coexists with childhood under-nutrition.

There are five major phases of fat-acquisition during human life, but only the first four have obvious benefits. The first comes in late fetal life, when we lay down a thin adipose layer over what was previously an almost fatless frame. We do this in anticipation of the metabolic disruption that will occur when we have to adapt to life in a chilly, demanding outside world. The second phase of fat deposition comes in early infancy, as we establish fat deposits to fuel rapid growth, intense activity and the construction of our huge brain – many of us are humorously chubby at the baby–toddler transition, regardless of our eventual shape. The third phase occurs at puberty in females, when oestrogens induce the fat deposits which give girls their distinctive curvy shape. The fourth phase comes during pregnancy and lactation, when adipose depots are enlarged by between two and five kilograms to meet the enormous demands of the growing fetus and newborn. (In fact, although the scientific literature does not mention it, I would insert another phase of fat deposition which occurs when we stop growing taller. Many women gain some weight after the age of eighteen, whereas men often stay adolescently skinny until twenty-one or so – because their bones stop growing later. The combination of copious alcohol and a no-longer-lengthening frame can lead to a sudden tendency to store excess calories.)

Each of these early four phases of fat deposition makes perfectly good sense, but it is the fifth, during middle age, which is less immediately explicable. (And we will come back to this shortly). We have already seen that many middle-aged people, in both developed and developing countries, do not put on weight, so weight gain is clearly not a universal phenomenon. However, the *tendency* to gain fat in mid-life if sufficient food is available is very clear, even if it is difficult to explain.

When humans get fat, it goes everywhere. If you lay down lard, much of it accumulates under the dermis of the skin, even in places like the fingers, feet and scalp – people who lose weight often have to buy smaller shoes and hats, and find that their rings fall off. However, fat is also deposited internally, inside the thorax and in the omentum, a sheet of membrane suspended from the stomach.

Unlike most species, humans exhibit dramatic differences in the amount and distribution of fat between the two sexes. At puberty, subtle pre-existing differences are amplified by the effects of oestrogens, and girls deposit more fat than boys – having between a third and a half more fat than males by adulthood. Most of this fat is subcutaneous and much of it resides in the breasts, hips and thighs, although women also have more subcutaneous fat in their extremities, too – even women's calves and forearms look curvier. Of course women do differ in the distribution of this new fat – some women have small thighs and large breasts, while others have large thighs and small breasts – but this shows that for each individual female teenager, fat deposition is an ordered, controlled process.

In contrast, men accumulate fat in a place which is far more efficient for running around on hunting and gathering missions – inside the abdomen and under the abdominal skin. Obviously a running fat man must work harder to move his fat in a forward direction, but at least he is not swinging it about at the end of his limbs. Imagine: if you were told to run a mile carrying lead weights, would you strap them around your calves or around your waist? In other words, belly fat is relatively efficient for locomotion although as we will see later, it may be particularly deleterious to health. The waxing and waning of male belly fat also has sartorial effects, as explained in a recent study commissioned by a major British clothes retailer. At the age of twelve, boys do up their trousers in the sensible place – around their waist. However, as trendiness and post-puberty muscle development kick in, the average waistband level drops to a buttock-exposing nadir at sixteen. Then commences a long, slow elevation of the trousers, passing the waist in the twenties and reaching a belly-squeezing zenith at fifty-seven. Finally, because most men lose weight thereafter, their trousers descend once more, back to the level of their anatomical waist, although the indistinct nature of older men's waists can allow trousers to descend rapidly and without warning to the level of the ankles.

These spectacular sex differences in body fat simply do not occur in other animals, and this raises the question of why humans have them. In many mammalian species, females are more prone to accumulate fat in times of excess, but they do not spontaneously accumulate it at puberty, nor does their fat distribution differ significantly from that of males. There are, for example, no curvy, Jayne-Mansfieldesque chimpanzees or gorillas, and perhaps it would be disturbing if there were – we all instinctively know that female curviness is a human preserve. There are many theories of why human females establish such distinctive fat deposits, and one of these is that our demanding system of reproduction, with overlapping offspring each trying to grow their own huge, fatty, energy-demanding brain, means that women must prepare for breeding in advance by storing fat. Add to this the fact that many of the more athletic aspects of food-acquisition may in our past have been conducted by men, with their central body mass, large stature and increased musculature, and suddenly an additional store of female fat starts to look like a valuable resource. Whether this trend was subsequently enhanced by men selecting curvy, fat-accumulating mates, I shall leave for my next book . . .

So why do so many people of both sexes accumulate fat in middle age? First of all, there are the usual suspects to consider. Middle-aged people tend to have less physically demanding jobs and take less recreational exercise than young adults. They also have more money with which to buy food and alcohol. They may also care less about their appearance, perhaps because they more often find themselves in stable sexual relationships. We all know couples who have piled on the pounds after finding true love, as well as individuals who have reacted to a split-up by trimming themselves down to a more predatory weight.

Yet being lazy, greedy and sexually complacent is not the whole story. Profound changes in body composition occur during middle age, and one of these is 'sarcopaenia'. Despite its alarming name, sarcopaenia simply means loss of muscle mass – and it is a change which is difficult to prevent. Although its effects are mainly noticed during middle age, and indeed it starts in the forties in women, it can start as early as the twenties in men. We are not entirely sure why sarcopaenia occurs, but it may be due to hormonal changes, degeneration of the nerves which activate muscles, or a combination of both. Obviously sarcopaenia has effects on strength and power – for example, hand grip weakens by fifteen per cent between the ages of forty-five and sixty-five. Considering what we said earlier about middle-aged men being the primary food providers in many human communities, it may seem strange that they experience muscular weakening at this time, but perhaps this goes to show that experience and guile are more important than brute force.

An important consequence of middle-aged sarcopaenia is its dramatic effects on metabolism. Muscle is an active tissue which burns a great deal of energy, so when it shrinks the body's requirement for calories decreases. Also, reduced muscle mass may have specific effects on the way we metabolize fats. The net result of these changes is that our rate of energy burning, the 'basal metabolic rate', decreases steadily over middle age, so that for every passing year each of us needs to eat ten fewer calories each day. Of course, this means we should eat less to maintain a stable weight, but often we do not find that easy. This in itself is sufficient to explain why middle-aged people who lose weight find it so hard to keep that weight off.

To make matters worse, middle-aged people do not realize they are gaining fat because that fat is often replacing the muscle they are losing. This 'straight swap' of muscle for fat means that their weight may not increase alarmingly, even though their percentage body fat is increasing fast. However, the muscle–fat swap does change the shape of the body. Muscle mass is mainly lost from the limbs (where most of the big muscles are) and fat gain is centred on the belly, in men at least, and this leads to the characteristic 'old man' body shape, with a fat belly and spindly legs. The moral of this story is that if you want to keep trim in middle-age, you should measure your waist rather than weigh yourself or calculate your body mass index.

There is controversy about whether women face different challenges. Anecdotally, many women believe they gain weight after their fertility wanes, not unlike the way neutered pets tend to put on weight. It would not be surprising if this were true because we know that the reproductive system uses an appreciable fraction of an animal's energy, so the cessation of reproduction would be expected to leave many calories unburnt. However, it has proved difficult to demonstrate that middle-aged female weight gain is caused by the menopause rather than by increasing age – the statistical link does not seem to be clear. Reversing some of the hormonal changes of the menopause with hormone replacement therapy does seem to correlate with smaller gains in belly fat, but other studies suggest that the menopause and weight gain are not directly linked. For example, women in high socio-economic groups often reach their maximum weight *before* the menopause even occurs. Also, women tend to exercise less after the menopause, and it could be this which leads to weight gain, rather than any direct effect.

Whatever the truth, fat distribution does change in middle-aged women, with the establishment of a more male-like, abdominal pattern of deposition – a pattern which women often dislike. Unfortunately, previous childbirth exacerbates these problems, as it tends to lead to increased waist, hip and thigh circumferences and general subcutaneous fatness, with an emphasis on concentration of fat towards the belly. As we have seen, however, such a 'central' location for fat is actually more efficient for movement so perhaps this trend, so often perceived as unattractive, does have its benefits.

So why does the body not have an inbuilt system for suppressing appetite to prevent mid-life lardiness? As it happens, the brain *does* control food intake, and it does so very well – indeed, it could be argued that the mid-life error of one gram per day actually shows just how accurate this control is. One element of this control system, the hormone leptin, was heralded as a route to preventing obesity almost as soon as it was discovered. Fat cells make leptin, and in rodents leptin suppresses the regions of the brain which drive appetite – so chubby mice eat less. Conversely, mice with a mutated, damaged leptin gene become extremely obese. At first sight it seemed that leptin might provide an excellent way to control human food intake – for example, reduced levels of leptin do stimulate appetite in humans. However, the data from humans soon became more confusing than those from mice. For a given amount of fat, middle-aged women make more leptin than men, yet this does not make them eat less and lose that fat. Also, adipose tissue in obese people actually makes *more* leptin than would be expected, yet it does not seem to reduce their appetite.

So it seems that we have been left with the worst of both worlds. Our hormones are good at making us eat more when we are skinny, but they do not make us eat less when we are fat. Our 'fat-o-stat' only works in one direction. Unlike mice, leptin does not seem to be an effective appetite suppressant in humans. Instead, leptin's main roles may be to maintain fat stores, and to drive fertility in women whose fat stores are sufficient to support pregnancy and lactation. Something in our past has made our bodies not worry about being fat.

*

And this is strange, because we know that obesity causes disease. Most alarmingly, it seems to be the 'central', abdominal pattern of fat distribution which is most damaging to our health in middle age. Perhaps 'obesity' is the wrong word to use, because we often take it to mean 'extremely overweight'. In fact, even slight increases in the middle-aged waistline are associated with an increased risk of disease.

The most important and interesting disease caused by middle-aged fatness is heart disease – it kills more people than any other disease, between a third and a half of all people, in fact. It has a worse prognosis than cancer (which has, contrary to popular belief, been partly 'cured' in the last few decades) and it also often runs a longer, more debilitating course. And most relevant to this book, heart disease breaks the rule that middle-aged people are, on the whole, a healthy bunch – forty per cent of heart attacks in the United States affect people aged between forty and sixty-five.

Yet, like middle-age itself, atherosclerosis, blocked coronary arteries and heart attacks are very much a human phenomenon – most mammals simply do not get them. Some great apes and birds get them, but the preponderance of cardiovascular disease in humans remains striking. We know that atherosclerosis has been with us for at least a few thousand years. For example, priests in Ancient Egypt often took home the meaty offerings left at their temples, and their embalmed corpses provide evidence that they often died young with blocked arteries as a result. Another disease often linked to obesity and heart disease also has a long history: type II diabetes. Indeed, this condition, in which an overweight body stops responding to insulin, and which encourages deposition of fat in coronary arteries, was described by the ancient Egyptians, Greeks and Romans. In fact, by the start of the eighteenth century, type II diabetes was (along with gout) one of only two human diseases linked to plenty.

The fatness-diabetes-heart-disease tangle is a complex one. Being fat directly puts greater strain on the heart, but it also damages the heart indirectly by causing diabetes, high blood pressure and disordered blood fat chemistry. In developed countries blood cholesterol and triglycerides increase between the ages of twenty and sixty, and blood pressure increases between thirty and sixty. Our pulse rate becomes more erratic in our fifties and the volume of blood pumped by the heart declines, too. And it seems to be the fat inside middle-aged bellies which is especially to blame, probably releasing chemical factors which make other organs ignore the effects of insulin. In comparison, subcutaneous fat is less damaging, so liposuction or a tummy tuck will not help your heart (and surgically extracted fat is soon replaced anyway, if dietary habits are not changed).

So why did humans evolve this ridiculous system, in which there is little restraint on our appetite, yet even slight obesity in middle age has such disastrous effects on our health?

One thing is certain. Middle-aged people no longer gather food, nor consume it, as they used to. Before the advent of agriculture, although our food supply was adequate, there was little incentive to put in the effort to acquire more food than was necessary. We did not eat much so we did not get fat. As a result, there was no need to develop hormonal systems to prevent obesity, because obesity did not happen. Later, with the coming of agriculture, most of the human diet was hard-won and vegetarian in nature. Famines became common, seasonal crops meant seasonal hunger, and social inequality within large human settlements made matters even worse. So for the last several thousand years, there was no evolutionary drive to prevent obesity, yet there was certainly a strong incentive to establish fat reserves in times of plenty.

Some anthropologists have suggested that the famine-adapted nature of human biology may go even further back than that – to a time hundreds of thousands or even millions of years ago, when climate change made food availability low and erratic. It is even suggested that much of our biology is that of a 'famine species': slow-developing, long-lived, big-brained to cope with adversity, and with low fertility which only increases when our females reach a certain fatness. In any case, the metabolism we have inherited is all about keeping the pounds on, rather than trimming them off.

This theory even has a name: the 'thrifty genotype'. The idea is that any species exposed to erratic food supply should evolve to respond to famine by ceasing reproductive activity and diverting resources towards simply staying alive. Give us food and we will store it rather than use it. And genetic variations in this in-built thriftiness could also explain why some sub-sections of the human species are more prone to obesity than others – Samoan emigrants to Hawaii, for example, or the variations in obesity in Caucasians of different ethnic origins across the United States (where, for example, obesity is more common in Hispanic women than non-Hispanic). Maybe some people have had to be more thrifty than others in the past. (Intensive selection for individuals who did not die of starvation or salt depletion during trans-oceanic slave transportation has also been suggested to explain the high rates of cardiovascular disease in African-Americans.) Perhaps thriftiness even explains why humans get atherosclerosis at all – maybe our ancient diets made it advantageous to be able to extract and mobilize every last smidgen of cholesterol and fat from our food.

An evolutionary perspective may also help us answer another great question of human life: why men do not live as long as women. This difference is partly explained by the fact that men are much more prone to cardiovascular disease, despite usually having less body fat. One reason for this is that we think that oestrogens have a 'heart-protecting' effect in women, especially before the menopause, whereas men have little circulating oestrogen. However, this finding does not offer a practical means to prevent male heart disease, because the toxic and feminizing side effects of oestrogens in men would be unacceptable. The menstrual cycle itself may also provide some protection from heart disease because the heart increases its pumping rate by up to a fifth at certain times in the menstrual cycle and also during pregnancy. In this way, the female heart is given a regular cardiac 'workout' every few weeks or years, and it is possible that this additional exercise makes it a fitter organ in the long run.

However, discerning the immediate causes of increased rates of heart disease in men does not explain why this unbalanced situation has evolved. One would think that natural selection would act to make men and women live for the same length of time. Indeed, in almost all mammals the two sexes do have a similar lifespan, which makes human males seem exceptional (although a similar discrepancy between the sexes may also exist in some whales, and as we will see later, this is not the only trait we share with our blubbery cousins). Again and again, men seem destined to die earlier than women, whether from heart disease, accidents, or drug and alcohol use. They even tend to neglect their own health: middle-aged men constantly rate their own health more highly than women do, even though their actual health may be worse. And we misguided health-optimists are also notoriously reluctant to seek medical help.

Surprisingly, there is an evolutionary theory which explains why men die young – why they keel over from heart attacks – and it is our old enemy, antagonistic pleiotropy. The theory predicts that genes which promote reproduction in early adulthood will flourish, even if those genes subsequently reduce longevity. And in human males, for whom attracting females and inter-male competition are both energetic, exhausting processes, a clutch of genes has appeared which help young men compete and breed, yet which harm them later on. The genes which produce testosterone are an excellent example. Testosterone drives male competitive behaviour and sexual activity in the young, but in the old it causes prostate and other cancers, while it offers no oestrogen-like protection to the heart. So, in short, men die young because they invest so much in their randy, stroppy youth. I can think of no clearer example of our prehistoric genetic history affecting a major aspect of modern human life. No wonder male heart disease is sometimes called the 'Flintstone diagnosis'.

We have come a long way in the first third of this book. Middle age has gained a context. Instead of being an irritating commonplace of human existence, it has emerged as a distinctive feature of the human life-plan, moulded by millions of years of natural selection into a form which may at times seem bizarre, but which we can now at least explain.

Yet how can I be positive about the fatness that so often comes with middle age? Well, now we know that even middle-aged fatness has been 'selected into us', just like the other four phases of fat accumulation. Because life was tough at many times during human history, middle-aged humans who are no longer breeding have acquired an amazing ability to be thrifty. Our metabolism is astoundingly efficient, and we should wonder at this great miracle just as much as we should worry about how efficiently it piles on the pounds when we eat too much. Obesity is a problem, but at the risk of stating the obvious, it *can* be controlled by reducing and changing our diet. And the fact remains that our ability to store and mobilize fat has powered most of human evolution – all those babies, all those brains, all that frenetic activity.

And middle-aged people are the thriftiest of all. The reason they put on weight is because they have become so energy-efficient. As reproductive activity declines, humans restructure themselves to burn less energy. Whether this is to promote their own survival or to liberate valuable food for their offspring we do not know, but *this* is the reason middle-aged people need so little food. Middle-aged fat metabolism is the ultimate human survival tool, and it is simply unfortunate that it cannot cope with our modern, unnatural superabundance of food – our 'obesogenic' environment. Fat has saved us so many times in the past, that we simply cannot cope with the fact that it is now more likely to kill us.

PART II

STILL CRAZY AFTER ALL THESE YEARS

The Triumph of the Middle-Aged Mind

What was astonishing to him was how people seemed to run out of their own being, run out of whatever the stuff was that made them into who they were, drained of themselves, turn into the sort of people they would once have felt sorry for.

Philip Roth, *American Pastoral*, 1997

7. Over the hill or prime of life?

Middle-aged people worry about their brains. They hear that a third of all people alive today in developed countries will develop dementia by the time they die, and they wait nervously for the signs. They see and feel their body changing – deteriorating, in their opinion – and fear the same will happen to their mind. They believe their brain will be their only 'saleable asset' in a few years' time, and the thought of that failing too frightens them. Yet perhaps they should worry less about the future and spend more time enjoying the present: in middle age the brain is indeed changing, but not necessarily for the worse. In fact, the evolution of the middle-aged human brain has been a success story.

Of course the brain is important – it has after all been the key to our species' success, which is why it gets the middle third of this book all to itself. Yet one thing is very noticeable about humans: they are not inherently very good at anything. They are not strong or fast or sturdy, nor are they born with a hard-wired ability to hunt, gather, speak or do. Unlike most animals, they are delivered into the world with very little programming, very little 'software'. However, they make up for this shortcoming by having amazing 'hardware': humans have a brain which is much larger than is usually required to operate an animal of our size, even a primate. And it seems that, given sufficient time, the brain can learn to do almost anything. It learns partly by practice and partly by being taught by other, more experienced human brains. Yet learning to do things can take us decades – we have already seen that learning to hunt may take more than twenty-five years – and this ability to learn a lot, but slowly, may be one reason why we humans live so long. But despite the time it takes, the human brain's flexibility, capacity and lust to learn transforms us from a feeble, inadequate species into the most powerful creatures on earth.

Having a brain like ours requires a lot of resources. Ounce-for-ounce, brains use a large amount of energy, and we humans have a disproportionate number of ounces. And as we will see, a distinctive feature of the human life-plan is that when we reach middle age this huge, demanding organ still functions well – it continues to burn energy and maintain itself in full working order even though it inhabits a body which is becoming sub-fertile and frayed around the edges. In other words, dramatic degeneration of the brain is uncommon in middle age. The extended maintenance of excellent brain function in humans demands explanation.

Of course, things do change in the middle-aged brain and I am personally very aware of this. My day job involves teaching Cambridge veterinary students between the ages of eighteen and twenty-one – often in intensive, interactive, small-group sessions. Most of them are brighter than me and all of them are quicker than me, yet I can usually stay a step or two ahead of them, and I know sixty-five-year-olds who regularly do the same thing. Of course we oldies know more than the students do, but this alone would not be enough if our thinking skills were deteriorating. To me it seems that we stay ahead not by thinking more, or harder, and certainly not more quickly, but by thinking *differently*. Every day I am made aware that my brain thinks in a different way from twenty years ago. Maybe it is compensating for something, or maybe it is simply improving – we shall see – but I believe everyone's ways of thinking change in similar ways in middle age. And this is because those changes are programmed into each of us as part of the developmental 'clock of life' which resolutely ticks onwards throughout our fifth and sixth decades.

So, in middle age, are we over the hill or are we in the prime of life? Is our brain developing or deteriorating? And why do some people's brains gradually start to fare so much better than others – causing so many of the unfairnesses of old age? As we will see, the human brain is an excellent example of why middle age is so interesting – it represents all the pay-offs, balances and compromises which have faced our species over the millennia. We try to maintain our ability to think, yet to do that we must change entirely *how* we think. Also, the middle-aged mind is interesting because it is subtle. We do not get obviously cleverer or stupider in middle age, but instead we change the mental means by which we achieve the same intellectual ends. This subtlety is, I think, why middle-age cognition was not studied much until recently – compared with the tumultuous cerebral transitions of child-hood, adolescence and old age, middle age can, superficially, seem like a period of dull stasis. But, as we are about to dis-cover, this is not the case.

Let us start with the senses. All information entering the middle-aged brain comes via the senses; there simply is no other input into the brain. The senses are a good place to begin because they are inherently easier to study than other aspects of the brain – you just expose people to stimuli and ask them what they perceive, or observe how they react. Also, the layout of the entire human brain is a direct reflection of the sensory information it gathers. The laws of physics mean that there are surprisingly few sources of information available to an animal – light, chemicals and movement, including sound vibrations. In fact, over the course of evolution, humans seem to have missed out on some senses – platypuses use electroreception to find food, and pigeons use magnetoreception to find their way home, but these abilities are rudimentary or absent in our own species. Thus we humans have built our brain around just a few key senses, and, unfortunately, evidence suggests that these few sensory inputs are weaker in middle age than when we are young.

Humans are among the most visual mammals in existence, what with our daytime, fruit-inspecting, prey-spearing ancient way of life. Because of this, it is visual decline we notice most. I first became aware that I could not focus on nearby objects when I was forty – the realization came suddenly one day when I was peering around the back of a computer to plug in a connector. The suddenness of my long-sightedness struck me as being entirely unlike a slow deteriorative process: instead it seemed as if it was 'meant' to happen – somehow preordained, or programmed to occur over a short period of time. Indeed, 'presbyopia', as it is called, has all the hallmarks of a controlled developmental process: it is rare at thirty-five, but universal at fifty. And its suddenness was one of the spurs to me writing this book.

Our ability to accommodate – to alter our eyes' focus to see near and far objects – peaks around the age of eight. My eight-year-old daughter often enthusiastically thrusts drawings in front of my face, only to look on in confusion as I rudely push them away to arm's length so I can make out what they are. We achieve accommodation by altering the shape of the lens which floats in the centre of each eyeball. However, our lens is a complex globoid of living transparent cells and it changes over time. Unlike artificial lenses, the extent to which the lens bends light varies throughout its thickness. In middle age the crystalline proteins in the central cells where light is deviated most, degenerate, clump together and become inflexible – and these changes occur regardless of how good or bad your eyesight was previously. This protein degeneration is hastened by heat, which may explain why presbyopia starts earlier in hot countries. The core of the lens loses its pliability and cannot be distorted as readily as before, and this rigidity spreads slowly outwards to affect the periphery of the lens, too. This, and changes in the arrangement of the fibres which suspend the lens within the eye (which are called the 'zonules of Zinn'), explains why we all need spectacles once we have run out of arm's length at which to hold our reading matter.

Hearing fares little better, and while only thirty-five per cent of people have significant hearing loss by the age of sixty-five, from childhood onwards we all slowly lose our ability to hear high-pitched sounds. Thus, this 'presbycusis' is not strictly speaking a feature of middle age alone, although many aspects of the ear do change during middle age. Probably the most important is that the sound-sensitive hair cells in the inner ear, or cochlea, deteriorate, but the eardrum and cochlear nerve degenerate, too, as may the sound-processing regions in the brain. Although some hearing loss is unavoidable during middle age, steps can be taken to reduce it, for example carefully managing diabetes, high blood pressure and atherosclerosis, as well as avoiding noise – so all that parental criticism of teenage clubbing and earphone use may have had a point to it after all. Also, smoking is known to damage middle-aged hearing by reducing blood flow to the ear, as well as by causing other diseases which adversely affect sound perception. In addition, for reasons which are not well understood, low socio-economic status predisposes to hearing loss even when other factors such as noisy occupations are discounted.

I am not sure 'presbyosmia' is a real word, but loss of smelling ability in middle age is a very real phenomenon, even if humans are not as consciously aware of smell as they are of the other senses – the brain region dedicated to smell is relatively tiny in humans compared to dogs, carp and even dinosaurs. It is much harder to quantify smelling ability, but it almost certainly declines after the age of fifty, if not before. For example, experiments show that the middle-aged brain exhibits lower responses to smelly stimuli than the brain of younger adults. And although we do not think about smell much, it is extremely important in many aspects of our lives, and declining olfactory ability may cause depression, loss of libido, anorexia and accidental ingestion of decomposing food. Also, our ability to taste chemicals on our tongue deteriorates as well ('presbygeusia'?) – women lose taste buds from their fifth decade onwards, whereas the decline is delayed for a further ten years in men, for some mysterious reason.

On the face of it, this decline in sensory ability seems alarming. It is easy to imagine that deterioration of the few sources of information coming into the brain might have serious consequences for its functioning, its very will to carry on, even. I can certainly see how this might be the case in the elderly, but apart from a few minor inconveniences, do many of us truly suffer because of sensory loss in middle age? Surely blossom, Bordeaux and Bach remain just as beautiful after forty?

I think there are three reasons why natural selection has allowed our senses to fail in middle age, and why we do not usually notice it. The first is that our brain does not slavishly examine every single piece of incoming information. The overwhelming majority of sensory information – the background hiss, the passing foliage, the smell of your own body – is discarded as irrelevant before it even enters consciousness. And very little of what does enter consciousness actually tells us anything useful about the outside world, so most of that is quickly forgotten, too. Thus, our brain is a triumph of information sifting, and because of this a partial loss of sensory acuity is unlikely to cause much of a problem. The second reason why we do not notice middle-age sensory loss is that our senses are, in reality, over-engineered. Our nose can discriminate chemicals which differ by a single atom; our eye can detect a flurry of less than ten photons; our ear can detect vibrations smaller than the diameter of an atom. This may sound amazing, but do you really need to be able to do it once you are forty? In fact, this leads us on to the third reason why middle-aged senses may deteriorate, which is that on the rare occasions when we need such staggering acuity, our social way of life means that there will always be a young person nearby to provide it. Hunter-gatherers usually prowl in mixed-age groups, and the fact that the young are often the first to hear the footfall of the prey does not detract from the middle-aged's ability to coordinate the actual kill. Similarly, is it worth maintaining the amazing sensitivity of human senses just so a hunter-gatherer does not have to ask their daughter if she can see a tiny discoloration on a piece of fruit? Let the young be our eyes and ears: middle-aged people are still the brains of the enterprise.

Studying the middle-age changes taking place within the brain itself is, unsurprisingly, more of a challenge. The brain thinks in many different ways and assessing these objectively requires a barrage of cognitive tests invented by psychologists over many years. Also, scientists love arguing about whether these tests are biased against certain people, whether individual tests actually test what they claim to test, and whether we are testing for the things we really want to know about. One thing is clear, though: there is no single test of the human ability to think, so the sensible approach to middle-aged cognition is to dissect it into its constituent parts.

Tests of different aspects of cognition give different results. Some tests suggest that our cognitive prime comes some time around the age of twenty and that we experience a steady decline thereafter. However, these tests are mainly those directed at *speed* of thinking – the ability to recognize items quickly, to make speedy decisions, and the ability to do just about anything under a tight time constraint. Younger adults react to time pressures simply by thinking very quickly and coming up with an answer, and this is something that middle-aged people find harder to do.

However, this apparent decline in speed-related cognitive ability is at odds with what we have already discovered about the contribution of middle-aged people to human life. Earlier, we saw that they are not just sedentary repositories of cultural information, but also exceptionally effective, active contributors to the resources of their social group. On the plains they probably hunt and gather more effectively than their younger peers and in the cities they earn more and have more political power. Yet how can this be reconciled with the discovery that they think more slowly?

In fact, these conflicting findings may readily be reconciled if we accept that absolute speed is not necessarily an important component of cognitive ability. Psychologists have been arguing about this for some time, but many now believe that speed is often not of the essence. As we have seen, many modern hunter-gatherer communities have a lot of time on their hands, and they positively relish protracted cogitation, recitation and debate. Admittedly, the final moments of a hunt probably involve some quick thinking, but I suspect that thirty years of experience may be more useful in that situation than a faster mind.

Bearing this in mind, it is perhaps unsurprising that a wide variety of cognitive tests, involving verbal skills, spatial perception, mathematics, reasoning and planning, are much more flattering to the middle-aged. Charted over the course of individual adult lives, many of these abilities trace the arc of a low, rounded hill – increasing from the twenties to reach a long smooth peak in middle age, and thence entering a slow but hastening decline. There are of course variations – maths ability peaks early, around forty, whereas verbal tests often peak later, around sixty. Remarkably, many abilities do not exhibit any statistically significant decline until after sixty-five.

This low, rounded hill of cognition is an important discovery. I firmly believe that many people worry about their cognitive abilities during middle age, and I think they could view these results in three different ways.

The first I shall call the 'summit euphoria' reaction, and it is the one to which I subscribe. Perhaps we should simply be pleased that, contrary to popular opinion, middle-aged people are at the peak of their intellectual abilities in many ways, and fairly uniformly so between forty and sixty. Far from being 'over the hill', they are in fact luxuriating in the sunshine at the hill's broad summit. Some earlier studies of cognition in middle age concentrated on reports, often anecdotal, of the unexpected ways middle-aged people sometimes solve problems, and viewed them as desperate attempts to stave off a cognitive decline which was already supposedly underway. However, we now have more recent results and they are clear: middle-aged people often 'think better' than everyone else. In other words, forget your crow's feet and revel in the fact that the middle-aged human brain is the most powerful, flexible thinking machine in the known universe.

The second perspective on middle-aged cognition I call the 'trouble ahead' view. After all, when you are at the top of a hill, the only way is down. Human beings are very good at thinking about the future – we may be the only creatures fully aware of the implications of ageing and death – and even in our hour of middle-aged cognitive triumph, our thoughts turn to the coming descent. Also, middle-aged people are in the perfect position to sense the first, fragmentary and often trivial signs of incipient mental decline. For example, because they hear so much about dementia in the media, or are so involved in caring for their own elderly parents, they become paranoid about memory, even though short-term memory probably does not start to deteriorate until at least the age of fifty. Long-term memory appears to fare even better, often surviving unscathed beyond middle age, and any 'senior moments' that do occur usually result from an occasional temporary failure to retrieve a memory, rather than the actual permanent loss of that memory – something that is perhaps more noticed by the diligent middle-aged mind, with its numerous competing responsibilities. Thus memory, although fretted about, is one of the most stable elements of the mind, which of course makes sense if middle-aged people are meant to act as transgenerational conveyers of human culture.

The third possible reaction to cresting the low, broad cognitive hill might be called the 'deceptive plateau' assumption. During middle age, the net sum of our cognitive abilities rises gradually to reach an almost imperceptible maximum and then coasts downwards equally slowly. Nothing more dramatic happens, and on the day-to-day or even year-to-year timescale, this can feel very much like stasis. Indeed, as I mentioned before, the relatively unchanging level of aggregate cognitive ability in middle age is what makes this period of life unattractive to researchers. Yet just because our *total* cognitive ability alters little, this does not necessarily mean that the *way* we think remains unchanged, as we will now see.

Within the last ten years or so, there has been one technical advance that has completely changed the way we can study the middle-aged brain, and that is Magnetic Resonance Imaging, or MRI. The principles behind MRI sound so unlikely that it is surprising that it ever works. An MRI machine consists of a huge magnet, so strong that it can align the spin of all the protons (mainly hydrogen atoms) in your head, and a radio transmitter which then knocks those protons all out of alignment. (I have had an MRI scan, and was rather disappointed that I could not feel any of this proton-malarkey going on inside me.) The protons in your head then snap back into alignment under the influence of the magnet, and in doing so transmit their own radio signal, which is then picked up by the machine and converted into a three-dimensional pixellated image of the inside of your head. Believe me, it does work.

With MRI we are now able to study both the structure and the activity of an unharmed, living human brain, and of course this is an immensely powerful way to follow how brains change over the human lifespan. Unfortunately, MRI has not been available long enough to scan individuals as they progress from thirty to seventy, so instead our middle-age imaging data are obtained by comparing scans from cohorts of people in different age brackets. This is not as powerful a method as following individuals throughout their lives, but we will have to wait a few decades to have those data.

Despite these limitations, it is already clear that there is a great deal of structural change going on inside the middle-aged brain. Let us begin by considering the grey matter: a dense tangle of nerve cell bodies and short connecting fibres, most of which reside in the corrugated surface layer of the two huge cerebral hemispheres. It appears that the grey matter steadily loses roughly a quarter of its volume between the ages of twenty and eighty. This may sound like a lot, but we should interpret this discovery with caution. First of all, the decline is apparently no faster or slower during middle age – so it is actually a feature of adulthood in general, rather than middle age in particular. Also, losing a quarter of your grey matter is not as spectacular a loss as it might sound, especially as reduced grey matter volume is not necessarily a bad thing. For example, grey matter volume declines precip-itously during adolescence when the brain is very obviously getting more adept at most cognitive tasks, and the adult decline may be little more than a protracted tailing-off of this process. In fact, grey matter volume often declines because unneeded or unused connections between nerve cells are being efficiently pruned away.

As with so many elements of middle-age change, the depletion of grey matter looks like a coordinated, structured process rather than a random deterioration. For example, grey-loss occurs in different locations at different times, yet the pattern seems consistent between individuals. Also, some studies suggest that the prefrontal cortex at the front of the brain, which is responsible for what we like to think of as our higher, 'executive' functions – planning, abstraction, complex intellectual tasks – loses grey relatively early.

Possibly linked to these changes in brain structure, we have now identified alterations in brain function, too, using novel imaging techniques which allow us to record activity in different regions of healthy human brains. In these studies, subjects are often asked to attempt one of a barrage of cognitive tests – of memory, recognition, naming, selection, and so on – which we believe assess the small mental building blocks that combine to create human sentience and intelligence. Certainly, they have shown that particular tasks are carried out in specific regions of the cerebral cortex, but they also show that the brain regions we use *change* as we get older. For example, there are variations within the prefrontal cortex: the regions further back (the dorsolateral prefrontal cortex) are relatively spared, while the regions nearer the front (the orbitofrontal cortex) lose volume sooner, and are also more frequently afflicted by the protein deposits seen in Alzheimer's disease. Indeed, some studies show that the ability to perform mental tests which require the activity of the orbitofrontal cortex declines sooner than for 'dorsolateral' tests. So while loss of grey matter is a normal, all-adulthood phenomenon, there are signs that it may still sometimes be related to cognitive loss – although we do not yet know if that is the case during healthy middle age.

Beneath the corrugated grey matter sheet lies a thick region consisting of interwoven nerve fibre bundles which connect nerve cells in widely separated areas of grey matter. This deeper region is called the white matter and it too has attracted considerable attention in imaging studies of the middle-aged brain. White matter connections are important, because a grey matter nerve cell is only as useful as the connections it makes to other cells – just as an isolated transistor can achieve little on its own. The brain is not amazing because it has lots of nerve cells: it is amazing because of how those cells are connected together.

The total volume of white matter in the brain seems to peak during middle age, and probably does not decline significantly until after the age of sixty. This, of course, matches the evidence that many cognitive abilities are also at their greatest in middle age. However, once again the pattern is not simple, with variations in the waxing and waning of white matter between different brain regions. For example, in some regions white matter declines steadily throughout middle age which bucks the general trend. Also, newer studies which claim to measure the organization and integrity of the white matter rather that just its crude volume, suggest that in some regions the structure of the white matter may be optimal in early middle age or even before. And variations in white matter integrity between different brain regions have been claimed to explain why middle-aged people lose the ability to think quickly or switch rapidly between mental tasks, long before their other abilities decline.

Studying specific pathways within the brain has also given us insights into what might be happening inside the middle-aged mind. It is notable that brain proteins which respond to the chemical dopamine (dopamine receptors) decline in number as we age, and the proportions of different types of dopamine receptors also alter, possibly as the amount of dopamine itself declines. This single change may be extremely important because the secretion of dopamine into the prefrontal cortex by deep brain structures is thought to be crucial in promoting cognitive speed, short-term memory, multi-tasking and keeping useful information 'in the front of one's mind'. Indeed, an increase in dopamine reaching the cortex is thought to drive the blossoming of mental abilities in teenagers. Although the converse, a dopamine decline, does not seem to cause any spectacular cognitive deterioration in middle age, it is entirely possible that the brain still must adapt to cope with it.

Imaging studies have given us the clearest evidence that the brain is adapting – changing the *way* it thinks in middle age. Although we can compare how middle-aged and young adults fare in cognitive tests, it is only when we peer inside the actual functioning brain that we discover that the two different age groups may use completely different brain processes to perform those tasks. The mental ends are the same but the brainy means differ.

For example, during many mental tests the activity of the prefrontal cortex is greater in middle age than it is in younger adults. Some have claimed that this is the sign of a flustered brain desperately compensating for its deteriorating abilities, but I would suggest that it is simply a change in the circuits which the brain uses to do things as it becomes more mature – and as a result sometimes performs better than it ever did in the past. In addition to its increased activity, the middle-aged cortex seems less inclined to concentrate certain activities in one hemisphere or the other. Preferential use of the right or left hemisphere is apparent in young adults as they perform tasks of recognition, memory retrieval or sequence manipulation, but is less marked in the middle-aged. And looking deeper into the detail of the cortex, brain imaging is now giving us ever-more detailed insights into how our usage of cortical sub-regions changes as we get older.

Thus, the brain-imagers have confirmed what the cognitive-testers always suspected, that the middle-aged brain does things fundamentally differently from the young adult brain, and that it consequently often does them better. I admit that we must always be very cautious about linking basic brain activity to cognitive performance in the complex world we inhabit, but the pixels on the MRI screen and the behaviour of the people around us do seem to match surprisingly well.

So the middle-aged brain truly is a triumph. Although the sensory information entering it is becoming ragged, and its internal processes do not work at quite the speed they once did, this does not seem to matter much. Quite simply, the middle-aged brain is at the height of its cognitive powers, and whether you subscribe to the 'summit euphoria', 'trouble ahead' or 'deceptive plateau' view of that feat probably depends on whether you are an optimist or a pessimist. One thing is clear, though: the genetic-developmental 'clock of life' still ticks inside the middle-aged brain, and it drives a radical restructuring of our thought processes, encouraging our brains to develop new ways of working well into our fifth and sixth decades. Over the next few chapters we will see what that means for each of us.

It makes sense for middle age to be a time of cognitive excellence – the brain allows middle age to be humans' most productive time, as well as a time when we are best able to convey our culture to others. Think of all the super-efficient hunting, gathering and cultural dissemination which our ancient forebears used to do, and compare it to the domination of the modern economic and political world by middle-aged people. The eras may be different, but the superiority remains the same, and it is mainly due to the middle-aged brain.

8. Why does time speed up as you get older?

I was forty years old when I started to write this book and *My! How the months have flown by!*

One thing middle-aged people think about a great deal is time. Of course, children are interested in the abstract concept of time and enjoy discussing how the days, weeks and seasons creep past. Young adults continually juggle and negotiate all the things they must do into the few hours which each day gives them. Yet in middle age our relationship with time takes on a new flavour, a new urgency – as we stand back and see how easily our precious time slips past, and we begin to wonder in a very immediate way just how much of it we have left. The fact that most of us can expect a good few decades more does not mean we can forget about time – if anything, it gives time even greater value. Middle-aged people are not destined soon to die, but the unpredictability of the future strikes home far more than when we are younger.

One feature of the subjective passage of time makes matters much worse in middle age: time taunts us by apparently speeding up as we get older. Weeks pass almost unnoticed and years roll past with an inexorably increasing swiftness. Can it really be May as I am writing this? It seems as if Christmas was only last week. Also, why is my younger daughter amazed when each autumn arrives, as if she can hardly remember the last one, so distant it hides in her memory? And in my seventies youth, did I, too, really fritter and waste the hours in an offhand way?

The phenomenon of accelerating time seems to be universal – all around the world people bemoan it, and it is even mentioned in ancient texts. When I ask my university students, they all agree that time does not crawl as ploddingly as it did when they were children, but eighteen-to-twenty-year-olds do not often consider themselves as being temporally short-changed. In contrast, in the fifth and sixth decades of life the shrinking length of each year starts to seem disconcerting, unfair, frightening even. So why does time speed up as we get older?

Before answering this question, I should probably start by admitting that nobody knows for sure why time speeds up. The study of subjective time has been an unfashionable detour in philosophy, history and science. It appears anecdotal, intangible and possibly intractable. But some thinkers have applied their minds and occasionally their experimental acumen to it, and I would like to present to you six theories which may explain the phenomenon.

Theory no. 1: The world speeds up, not you

It has been suggested that the acceleration of time experienced with age is a result of external world and cultural events occurring at an ever-quicker rate. We like to think (correctly or not) that life was paced and constant twelve thousand years ago – that each pre-agricultural generation took on the mantle of that which preceded it and continued to live in the same old way. Maybe the human way of life changed slowly, as environmental conditions tilted and shifted, but that change was imperceptible to individuals who never lived longer than a century. Then, with the advent of agriculture, the rate of change of human life increased, with settlement, property, writing and empires appearing in the space of a few thousand years. And thus began a gradually accelerating accumulation of cultural and technological innovation that has left us where we are today – call it progress, if you wish. And this process has indeed accelerated – scientific advances in the last decade have been faster than ever; technology has allowed us to communicate in ways which would once have been called electronic telepathy; the worlds of art and literature now change so fast that there is rarely time for an appreciable body of work to be allocated into any of the '-isms' which pepper the earlier history of the creative arts.

Yes, human life is changing exponentially faster; but is an increasing disconnect between middle-aged people and present day culture sufficient to explain the subjective acceleration of time? I suspect not. For a start, there is little evidence that the phenomenon has become more marked as human cultural change has accelerated – the ancient Greeks seemed to feel it just as keenly as us, although admittedly the sorts of ancient Greek who wrote about this stuff may have had a greater sense of hastening 'intellectual progress' than most. However, the fact that my students – who are at the perfect age to keep pace with contemporary culture and technology – also experience the subjective speeding-up, suggests to me that culture is not behind it. Changing culture may slightly exacerbate the subjective impact of accelerating time on the middle-aged, but it is unlikely to be the main cause.

Theory no. 2: It is all about how much time we think has passed

Since the late nineteenth century at least, thinkers have attempted to calculate the actual rate at which time speeds up with age.

One suggestion is that, for each of us, the speed at which time seems to pass is in direct proportion to the amount of time that has elapsed since we started to remember things long-term – at perhaps the age of three. In other words, the more memories you accumulate, the faster time rushes past. This is not so much a *reason* why time goes faster, but more an observation. It may seem a neat idea, but is has its problems. For a start, it would mean that at the instant our first memory is formed, time should be passing infinitely slowly – you have no memories, so time cannot pass – and I am not sure that all three-year-olds experience a frozen moment of eternal numinous insight. Second, I am certain that children form many memories long before the first memory which survives to adulthood, but those other memories simply do not last. Do they count in this calculation?

An extension of this theory is that the speed at which time seems to pass is in proportion to the total amount of *subjective* time that has elapsed since infancy (rather than the *actual* amount of time). This is, I admit, a confusing idea, and you may wish to pause and think about it: it means that the passage of one year of 'childhood time' would have a greater hastening effect on the subsequent speed of time's passing than the passing of one year of 'adult time' – because 'childhood time' seems to last longer. This theory may seem convoluted and tangled – subjective speed of time depending on previous accumulation of subjective time – but mathematicians have a way of coping with such things and have generated a formula for the subjective speed of time at any chronological age.

Yet just because we can make some maths from a theory does not mean it is correct. The mathematical model still tells us nothing about *why* the rate at which time appears to pass should be related to our accumulated experience – objective, subjective or otherwise. At best it is an attempt to describe the phenomenon rather than explain it. And it does not even seem to describe it very well. While some studies involving questionnaires about perceived distance of past events fit the predictions of this theory quite well, other studies which instead focused on perception of time intervals in the past do not fit the theory as well.

Theory no. 3: We distort time to stop us worrying

This theory relates to how we may manipulate our sense of time to improve our mental well-being, or alternatively succumb to anxiety about impending death.

When we reach middle age, we often feel we have reached a crossroads in our life, and one feature of this life-stage is that it is, indeed, central. We reach forty and we realize that we are probably halfway through our lives (or a third of the way through, as I claimed with characteristic over-optimism at my own fortieth birthday). No matter how positive I am about middle age, I cannot ignore the fact that people over forty are always nearer to death than they were at birth. Is it this sudden realization of 'time left' which makes time seem more valuable? And is our panicked response to this realization to watch every precious year slip by with greater attention, fear, and thus subjective speed?

In fact, it has proved possible to measure people's perception of death. A popular method is to present volunteers with a straight line and tell them that the left-hand end of the line represents their birth and the right-hand end their death. They are then asked to mark on the line where they think they currently stand. If the expected right-hand 'death point' is then calculated using the sort of information used by life insurance companies, some interesting results appear. For example, women are much more accurate at this test than men, who tend to believe they are further from death than is really the case. Also, as both sexes get older we start to believe that we are further from death than we really are. Other studies even suggest that people actively manage their perception of life to avoid excessive worry about death – we seem to push negative memories far back into the past and push expected negative events further into our future than they really are.

So could active management of the perceived time-frame of our life explain why time speeds up? A big problem with this theory is that it does not tie in with people's expressed feelings. Some people worry about death more than others, but their anxiety about death does not seem to correlate well with the extent to which they believe time has accelerated. People fear death: time speeds up. Yet the two processes appear to be independent. Death-anxiety may make time-acceleration more frightening, but it does not cause it.

Theory no. 4: Our memories are distorted, and this distorts time

We all know that memory plays tricks on us, so if memory is an important part of how we perceive time, could this explain why time plays tricks on us, too?

Most of us remember recent events more clearly than

events further in the past. We have an inherently better recall of things that have just happened, and a clearer sense of the order in which they happened and how they relate to other recent events. Thus, for the last year or so we each have a neat temporal framework in which our life took place – everything fits together. If we think back any further than that, events become progressively more fragmented and disconnected. We can remember them, but we must often resort to tricks and aide-memoires to work out when they occurred in relation to one another. For example, I use major yardsticks such as births of children and house moves to calibrate the relative timings of more minor events which occurred between twenty and five years ago. I retain vivid memories of those minor events, but I have lost the clear temporal stream in which they took place.

It has been claimed that this lost sense of time underlies an *illusion* that time is passing more quickly. Recent time is undeniably more ordered and catalogued, but this theory suggests that this coherence makes it seem to have occurred more rapidly. Is recent time subjectively 'compressed' into a shorter time-frame simply because it is neat, structured and fresh? Conversely, does the brain interpret the disordered temporal structure of our more distant past as a sign that it stretched over a longer period?

Life may be even more complex that that, however, because our sense of the timing of past events appears to change with age. Nowhere is this more clear than in the subjective recall of external, world events. If middle-aged people are asked to rapidly estimate how long ago an event took place, they consistently underestimate the passage of time. Of course, they then compensate for this error by calculating more precisely by reference to major life-event yardsticks –

and are frequently horrified by how wrong their initial esti-mate was. A personal example is how constantly surprised I am by how much of the music to which I listen, and which to me seems so familiar and contemporary, was in fact recorded over twenty years ago. (Maybe this is why I still set the boundary between 'old' popular music and 'new' popular music somewhere around 1975, when I was six.) In contrast, elderly people do the opposite – they place external events too far in the past, as if to put them in some long-lost youth-ful idyll. Perhaps middle-aged people are less keen to distance themselves from that idyll.

Whatever you might think of this theory, it is certainly rea-sonable to believe that we manipulate our memories and that they manipulate us, and that memory is a major part of defining the perceived course of our lives.

Theory no. 5: Less new stuff happens

The fifth theory of why time speeds up relates to novelty. It is common experience that doing something for the first time seems to take longer than doing it a subsequent time. The first day at a new school or a new job, or even the first day of a holiday, seem to crawl past. Also, the striking and frighten-ing novelty of a personal accident may stretch time even further, and people frequently report dramatic subjective expansion of fractions of seconds during which their life is in danger.

Novelty-induced time expansion is a well-characterized phenomenon which can be investigated under laboratory conditions. Simply asking people to estimate the length of time they are exposed to a train of stimuli shows that novel stimuli simply seem to last longer than repetitive or unre-markable ones. In fact, just being the first stimulus in a moderately repetitive series appears to be sufficient to induce subjective time-expansion. Of course, it is easy to think of reasons why our brain has evolved to work like this – presumably novel and exotic stimuli require more thought and consideration than familiar ones, so it makes sense for the brain to allocate them more subjective time. In contrast, the mundane and the predictable are permitted to trundle past in a barely acknowledged procession.

And when we are young, everything is novel. When you are a child, so many of the things you experience are being experienced for the first time. Even if they are not entirely new, then you are still concentrating on developing inventive new responses to them. So it has been proposed that time speeds up as we age because we encounter new things less often – our brain's response to the predictable world of middle age is to hurry us through it. Early life is stretched and expanded by novel challenges, whereas mid-life collapses before our eyes, deflated of fresh experiences.

So why does novelty affect our subjective perception of time passing? Perhaps this is all about memory: maybe the need to lay down fewer memories in middle age makes our brain believe that less has happened, and thus that less time has passed. Alternatively, it is possible that the link between novelty and the brain's internal perception of time is a very direct one. Perhaps time perception is actually *dictated* by external novelty – new experiences may be the very thing which makes time seem to pass.

Yet there are complications to these novelty-based theories. For example, many people report that childhood and adolescent experiences are intrinsically more vivid and more protracted-seeming than experiences in later life. Yet both age groups, and especially teenagers, constantly complain of being bored. We all know that time passes incredibly slowly when we are bored or waiting for something, yet it does not make sense to argue that adolescent life passes slowly because it is simultaneously novel *and* boring – surely novelty and boredom are opposing concepts? And boredom's effects on subjective time are confusing, anyway: a week of boredom passes painfully slowly, yet when recalled it seems almost to vanish in our memory because there was so little contained within it.

Another factor which may be important is enjoyment, and perhaps this is a better conceptual opposite to boredom than novelty is. We all know that time flies when we are having fun, yet busy, enjoyable weeks expand in the memory relative to the weeks of monotony which surround them. They appear to have taken a long time to pass, even though they slipped by so fast at the time. Thus, the subjective perception of time seems to depend on a mixture of novelty, boredom and fun; also, the sense of how time passes while we are doing something does not always agree with how much time, in ret-rospect, we felt we spent doing it.

Indeed, the nature of the attention we give to time may be one of the things which changes as we enter middle age. As middle-aged people's children grow up and their careers settle down, they have more time to ponder the passing of time on a 'macro' scale – they can sit back and worry about time trickling away. Yet middle-aged lives are often more con-trolled by routine and necessity than those of the young, so perhaps middle-aged people have less opportunity to actively consider their use of time on a 'micro' scale. Is it thus the curse of middle age to occasionally look back and be sad-dened by how we ignore time as it passes?

Time now, time then, new things, fun things, boring

things. Subjective time is looking increasingly fragmentary. Is there, in fact, any such thing as a single clock of time within each human head?

Theory no. 6: We have lots of clocks, but few of them are any good

This theory is where most of the science is going on, and it involves the study of the clocks in our heads.

There is good evidence that our brains contain clocks, and at least one of them is very accurate. There is a clock circuit of nerve cells in the lower part of the brain, called the suprachiasmatic nucleus, which ticks out fairly regular twenty-four-hour intervals – it is a major reason why our activity and sleep patterns follow a daily cycle. For example, there exists a mutant strain of hamster with an alteration in one gene used in the suprachiasmatic nucleus, and hamsters with two altered copies of the gene exhibit twenty-hour activity cycles (and, charmingly, hamsters which inherit a mutant copy from just one parent think the day is twenty-two hours long). The suprachiasmatic clock can be astoundingly accurate, and many people are struck by their ability to train themselves to wake up just a few seconds before the morning ringing of their alarm clock. Yet despite the fact that this daily clock is the most accurate internal timer we possess, our need for it is limited – after all, most humans throughout history could just as well have used the rising and setting of the sun to tell them when to wake up and when to sleep.

Leaving the daily clock behind, mental time measurement on timescales shorter or longer than a day – intervals for which no external day/night-type corroboration exists – seems to be a more error-prone and mysterious process. One theory of how we measure these time spans holds that we have other ticking clocks in our brain, akin to the daily clock of the suprachiasmatic nucleus, and each consists of three components: a 'ticker', a counter of those ticks, and a memory of previously experienced time intervals with which to compare that count. An alternative theory suggests that we do not have specific subjective clocks, but rather that our perception of time is a more general thing – a cumulative measurement of how much the brain has done or how much information it has received within a given interval.

Of course, this latter model could explain why time seems to expand or contract when we are excited or bored, but it could also have specific implications for middle-aged time-keeping. For example, we saw earlier that less information comes into the brain as the senses dull in middle age, so could this make each day seem less full and thus in retrospect appear to pass more quickly? We saw also that the body's metabolic rate decreases in middle age – so could a slowing of the brain's activity make the outside world seem to pass more rapidly in comparison? However, both of these effects could be cancelled out by the fact that middle-aged people sleep less than younger people, so with more waking hours they could potentially end up with *more* sensation and *more* thinking taking place each day.

These speculations are all very well, but what do we discover when we experimentally assess how people of different ages measure time? Unfortunately, there is a problem with this approach: as it turns out, there may be more than one way to estimate time.

First of all, you can ask people to look back and compare time intervals, and also ask them to estimate how long, in seconds or minutes, time periods were. A crucial aspect of this sort of experiment is that the subjects should not know what is required of them until after they have experienced the intervals – they are kept in the dark about the nature of the experiment until asked, subsequently, to make their time estimate. Although age is not as important a factor as how exciting or boring these 'retrospective intervals' are, it seems that middle-aged people give significant overestimates of past time intervals when compared to younger adults. This is intriguing because it is our first tangible evidence that middle-aged brain-clocks run at different rates from young-adult ones. However, this result does not really fit the time-speeding-up-as-you-get-older phenomenon, because surely that would require middle-aged people to perceive that events in the recent past rattled by *faster*?

The second way to study people's estimation of time is to ask them to reproduce – to tap out with their finger – a time interval immediately after they have been exposed to it. Compared to women, men tend to tap out intervals which are too short, and it is not known why they are in such a subjective rush. Also, people tap out intervals which are too long as they get older, and it is intriguing to consider what this might mean. Could it be that our small-scale timekeeping deteriorates as we age? Indeed, there is evidence that time estimation deteriorates, perhaps even before middle age, and that the extent to which it deteriorates in individual people may parallel alterations in the size of some prefrontal brain regions.

The third way to assess timekeeping is to ask volunteers to 'produce' intervals – to confiscate their watch and then ask them to count out time periods of a certain number of seconds. When this is done, the results are striking: in one study which invited participants to produce intervals between ten and three hundred seconds, there was a signifi-cant acceleration in timekeeping with age. On average, twenty-year-olds get their predictions pretty much correct, but accuracy deteriorates during middle age so that by the age of sixty, people shave off roughly thirty per cent of the time intervals. This is a fascinating result, but surely it is once again counterintuitive – if middle-aged people's internal clocks are speeding up, should the outside world not seem to be flowing by more slowly, instead of more quickly?

Things become even more confusing when different time spans are measured. A study of the production of intervals between one and twenty seconds showed no such speeding up in middle-aged people, leading to the suspicion that there may be fundamentally different processes involved in count-ing below-ten-second intervals and over-ten-second intervals. And indeed, we also now have good evidence that assessment of time over different timescales may involve activity in different regions of the living brain. So if we have different clocks for timescales as similar as these, how can we possibly extrapolate these results to help us understand middle-aged people's subjective assessment of how the *months and years* of their lives have passed?

As I admitted earlier, we still do not know why time passes more quickly in middle age, but this does not stop the phe-nomenon being the fascinating, but unspoken, backdrop of a large part of our lives. After all, our lives take place in an unstoppably hastening torrent of forward-flowing time.

As we have seen, experimental data hint that we may use different types of timekeeping for different purposes – differ-ent clocks for within-day, daily and supra-day time estimation; one clock for time passing as events actually happen, and another clock for looking back on those events with hindsight. Maybe this explains the inconsistencies in what middle-aged people sometimes say about time – the days may drag on while the years scream past, yet suddenly everything seems so long ago.

But one thing is clear. With the exception of the twenty-four-hour suprachiasmatic clock, whichever clock we study, middle-aged people are not very good at measuring time. And to me this suggests something important: the reason middle-aged people are not good at small-scale (and perhaps long-term) time estimation is that it has not been an important skill in their evolutionary past. The natural world is replete with organisms which predict days, lunar cycles, tides, years and even prime number multiples of years with bewildering accuracy, and yet middle-aged people cannot even standardize their own perceptions of time. Presumably, for the thousands of generations of campfire hunter-gatherers which preceded us, the important decisions like knowing when to spear the antelope or when to flirt with a prospective partner did not involve counting out arbitrary subdivisions of time. And because of this, nature never made middle-aged people very good at it.

However, the subjective acceleration of life during middle age remains a remarkably consistent phenomenon which still requires explanation. Why has natural selection favoured humans who feel that life is whizzing past in their middle years? What could be the possible evolutionary advantage of this unnerving feeling?

In answer to those questions I would like to offer a specu-lation. When humans are children and teenagers, their time is mapped out by their own development and their urgent fight for survival. Until at least ten thousand years ago, parenthood ensued soon after, and the rigid temporal tyranny of child-care took over from that of growing up. However, when humans reach middle age, their life becomes less dominated by the need to continually produce little parcels of food or affection for their offspring, and their time, relatively abruptly, becomes their own. They have new, considered choices to make about improving the lot of themselves, their kin and their clansmen. The spur which drives them to make those choices is a new feeling that life is finite, as is their potential future contribution to human society. I propose that time speeds up to give those choices an urgency – to make us consciously stand back and consider our future lives. But not to stand back too long, of course, for there is a great deal of productive life still to be lived.

And some would say that mental phenomena as distinctive and subtle as the ways in which middle-aged humans view the structure and meaning of their own lives could not evolve – that they could not be hewn by the brutal hand of natural selection. However, in the next few chapters I will argue that this is exactly what has happened.

9. Is your mind 'complete' by the time you're forty?

Until recently, many psychologists did not think much of note happened inside the middle-aged mind. From the time of Sigmund Freud's doctrine that it is the early phases of life which are crucial to the development of the mind, middle age was viewed as a period of relative stasis, during which little changes. The very idea of 'adult development' was a contradiction in terms, and many later psychologists continued to assert that human personality does not alter during middle age. Once you reach adulthood you are finished, complete – or so the story went.

However, sitting uneasily alongside this assumption of middle-aged stasis was a suspicion that middle age brings with it a new round of stress, sadness and mental illness. Many of us worry about this as we approach middle age. It is possible that such an avalanche of woe, if it exists, could potentially be explained away as a sign of a static, unchanging mind failing to cope with all the stresses which come with a new life-phase. But if we make two simple assumptions about people – that the mind is the central element of a human being, and that middle-aged humans are products of aeons of natural selection – then we are left with an uncomfortable question. Why would evolution leave middle-aged people with minds so immutable that they cannot cope with simply getting older? An unchanging middle-aged mind does not make sense.

We will now extend our exploration of the middle-aged mind by challenging some of the assertions commonly made about it. In this chapter, let us focus on the question of whether the mind and the personality really do change over the fifth and sixth decades of life. We will also consider whether it is change or stasis which explains two common stereotypes of middle age – frustration at lack of control over our lives, and increasing social and political conservatism.

Although Freud's ideas remained dominant, it was not long before a few psychologists started to question the idea of the static middle-aged mind. Having a big name behind a theory can count for a lot in psychology, so it is important to mention that Karl Jung's ideas on the development of personality differed dramatically from Freud's. Indeed, Jung maintained that it was unlikely that the human personality would be mature even by the age of forty or fifty. Supporters of this view have subsequently claimed that even crude arithmetic supports the idea of middle-aged change. Put simply, a middle-aged mind has had roughly three times longer than an adolescent mind to become different from other personalities – longer to veer away from the influence and insistence of others and acquire its own tendencies, characteristics and foibles. Furthermore, the lives which people lead after the age of eighteen differ more from each other than the lives they led when they were regimented in the parental home and at school – adult life is inherently more able to pull people's minds in different directions.

So the simple fact that middle-aged people have had more time, and more heterogeneous time at that, should, in theory, make them more *individual*. Indeed, many psychologists claim that this is exactly what happens to the human experience of life in middle age. Whereas young adults can share the same processes of thinking, planning and hoping for the future, by middle age that future has become a real and manifest present. Those of us who have done well and coped well relish the multiplicity of new challenges brought by middle age, and recount positive, affirming stories of our own lives. In contrast, for those whose thoughts, plans and hopes came to little, middle age can be a time of failure, anger and frustration. This is not mere speculation – it is the message that emerges from carefully planned psychological studies. Once again middle age appears as the central point of our lives – the time when youthful aspirations crash into mature reality, a realization as yet not softened by old-age stoicism.

More recently, experimental studies have confirmed that middle age is not a time of psychological impasse, yet neither is it a time when all our psychological cards are recklessly thrown into the air to fall into an entirely new pattern. Instead, our previous personality is progressively but only partially modified as we age. Because of this, positive, healthy people tend to stay positive and healthy while negative, unhealthy ones stay negative and unhealthy. For example, it is striking how teachers' assessments of children's personalities can be used to predict those pupils' future middle-age mental and physical health, weight, and use of alcohol and nicotine.

Yet superimposed on this tendency for psychological continuity are some consistent changes in the middle-aged world-view. Middle-aged people tend to worry more about losing control of their lives in the future, even though middle age is actually the time when many people have greatest social and economic power. Also, over the course of middle age, people start to react more conservatively to new situations – increasingly they strive to prevent bad things happening rather than to increase the chances of good things happening. According to various studies, middle-aged people also grow more certain about their identity, become more conscientious, more 'agreeable', seem more eager to spread their efforts over many activities, and are keener to help younger people.

Some of these alterations probably result from changes going on in the world in which the middle-aged mind finds itself – the inexorable growing-up of children, the vicissitudes of a career, the undeniable bodily changes. However, the changes to the personality are so consistent and distinctive (some might say clichéd) that it seems as though their essence is built into the human developmental programme. Yes – I believe we each carry in our genetic instruction manual elements which can even change the way we think about ourselves as we get older. This is why people with no children, an unchanging career and a fortuitously unblemished body *still* change how they think when they reach middle age. You might not like to believe that something as self-defining and complex – something that is so much *you* – as your personality could be under the control of those little stringy genes in your cells, but the sheer consistency of middle-age psychological change certainly looks like it is controlled by something intrinsic to all of us. I would argue that the way we think at different times of our life is largely guided by the genes we inherit – our own personal version of the 'clock of life'.

Whether they agree with such outrageous 'genetic determinism' or not, most psychologists now believe that human minds do indeed continue to develop long into middle age. In fact, it now seems strange to think back to a time when child and adult minds were considered utterly different – one which develops and one which does not. When measuring psychological development during any period in life, we no longer think in terms of a single linear narrative in which everyone's personality progresses through exactly the same preordained, sequential stages. Indeed, we now think the only reason children's personalities sometimes appear to develop in such a predictable and uniform way is that we force them to pass through fixed educational stages at certain chronological ages. In the absence of such common and clearly structured regimentation and externally imposed rites of passage, middle-aged psychological development is a symphony of many different but concurrent changes – some fast, some slow, some continual, some intermittent. And although the exact details of all those little changes will vary between individuals, different people's symphonies still sound similar enough to suggest that their central themes are products of our gene-conveyed shared evolutionary heritage.

One of the prominent melodies in the symphony of middle-aged psychological development is 'control'. Control can mean several different things, but middle-aged people do worry about control a great deal, even though they wield so much social, economic and political control over everyone else. Nowadays, middle-aged people often seem to run the show, as the average age of politicians, factory foremen, managers and professors has crept downwards into the fifties and even forties – one does not have to look far back in history to find a time when Britain, for example, was run by men who at least *looked* strikingly older than the people (still usually men) who run it today.

Yet not everyone is a captain of industry or government minister. Another form of control is one to which we can all lay some claim, and that is the control we each feel we have over our own environment. In the same way that many behaviourists believe an animal's welfare may be measured by the control it can exert over its environment – its perceived ability to avoid pain and want, or seek resources or comfort – many psychologists also believe that human well-being is largely dependent on pretty much the same thing. We all know that some things in life are eminently decidable by ourselves, whereas other events are entirely beyond our control – but what seems to be important is the extent to which we *believe* we can control the world around us. In its most crude form, this idea of control divides people into two groups – those who believe that everything in life is amenable to their own control, and those who believe that all they can do is react passively to whatever life throws at them. We probably develop strongly held ideas about this sort of thing when we are young, presumably based on our genes and our experiences, yet by the time we reach middle age we have had enough experience of life to ensure that our beliefs about how much control we have are even more deeply etched into our personalities.

Whether or not we believe we control the outside world is important for our general welfare. People who believe that control lies within them tend to be higher achievers, more motivated, less anxious, more able to cope with adversity, physically healthier and, apparently, happier. Although constantly believing that you can shape your own future can put pressure on you, and even be exhausting and just plain wrong at times, it seems that the upside of this blind faith in your own power is extremely beneficial. For example, it is claimed that a sense of self-determination drives a flood of brain transmitter chemicals which promote positive moods and physical health. And the good news is that most psychological studies of 'feelings of control' over the lifespan suggest that we feel most in control during middle age. In fact, rather as we saw with cognition in Chapter 7, there is a long low hill of perceived control from which middle-aged people can gaze down at the younger and older unfortunates around them. Unlike younger and older adults, middle-aged people often cite their career and caring for others as the two most important things in their lives, and it appears that they also consider these aspects of their lives the most controllable. There even seems to be a virtuous circle in play here, too: it is thought that a strong sense of control promotes success at work and good parental relationships, and conversely, success in these two key areas reinforces our sense of control.

However, all is not perfect with our sense of control during middle age. Although *average* ratings in psychologists' tests are higher during middle age, this does not mean that every middle-aged person believes they are the potent master of their own destiny. For everyone who finds that the world bends to their every whim, there is someone who feels their life is chaotic and unpredictable. For example, white collar workers are more likely to feel that they control their lives than blue collar workers. Also, educational background seems to have an effect, too. And women, in general, are more likely to feel that the world 'happens to them', rather than they 'happen to the world'. Middle-aged women also sense that they have greatest control of elements of their lives which involve interacting with other people, whereas men feel they are better at controlling 'things' – objects, money or abstract entities. Of course, many of these perceptions may reflect bitter experience. Maybe middle-aged manual labourers develop a sense that the world is beyond their control because the story of their own life shows that much of it actually *is*. Also, it is intriguing to speculate why women might feel the same way: do women have an inbuilt tendency to feel impotent, or is it their own experiences which make them feel that way?

One of the most important forms of control we have is the power to regulate and manage our own identity. Each of us has our own discrete, unique identity – a summation of what we think of ourselves and where we stand in society – but psychologists do not believe that this identity is unchanging. Far from it: although our identity is crucial, it is thought to be in a state of continual flux, forever recasting itself in response to the changing environment and our own personal attitudes. And here people differ. Once again, oversimplifying the human race into two extreme camps, some of us have a much more resilient identity and try to mould the world to fit us, whereas others have an incoherent, malleable identity which continually changes to adapt to the prevailing conditions in which it finds itself. Having a more resilient identity has sometimes been posited as the very definition of successful ageing, but it does have its problems. People with this sort of identity may be confident and optimistic, but they may also ignore the very real changes taking place in their lives, and find it harder to adapt to necessary change. Conversely, having an excessively labile identity also has its disadvantages – it can allow the ageing process to lead to panic, or even premature surrender to senile deterioration.

Yet psychological studies tell us that the general trend over middle age is for all our identities to shift from the resilient end of the spectrum towards the labile end. The end-result may be more noticeable for women because their identities are, on average, more labile to start with. This may sound worrying, but there is a great deal to be said for altering your sense of yourself during middle age. After all, the intense youthful drive to assert one's individuality seems redundant or even ridiculous in middle age, by which time you either have, or have not, achieved what you set out to do. Studies show that in middle age people's goals become less rigid, their aims more flexible, and this of course increases their chances of realizing those aims and goals. Maybe once we reach middle age, it is time for a little give-and-take when it comes to dealing with ourselves and the world around us.

A second major theme of the middle-age psychological symphony is conservatism – both social and political. We all laugh at the stereotype of middle-aged people becoming grumpy and reactionary, but does this actually occur, and if so, could it really be a product of evolution?

Middle-aged people exhibit distinctive patterns of socialization. Studies show that they engage in social interactions less often than younger people, and this seems superficially to fit the cliché of the middle-aged misanthrope. However, in questionnaire studies, older adults rate their social interactions more favourably than younger people, and this includes their dealings with their spouse and family. In fact, they actually seem to *increase* social interactions with a small clique of people – many middle-aged people, for example, renew their friendships with their brothers and sisters, often after years of distance caused by unresolved sibling rivalry or the all-consuming tyrannies of career and childcare.

Thus, it has been suggested that people do not become less sociable in middle age, but rather they are more selective about whom they socialize with – concentrating their social energy on a chosen, important few. Some have even argued that this change corresponds to certain theories of the human life-plan. They propose that when humans are young, and their time on earth seems to stretch far into the future, their interactions with others are based on acquiring information – that all-important human currency. As they enter middle age, time suddenly starts to feel like a finite resource (perhaps for the reasons discussed earlier), and their social focus changes from gathering knowledge towards generating emotional links with others. Of course, this chimes with the evolutionary idea that human life is partly organized around the process of conveying cultural information from the old to the young – in this case, the psychological studies have given us evidence of young people desperate to learn from whomever they can, and middle-aged people keen to form supportive bonds with close family and friends. This is a consistent change, a *development*, in the way middle-aged people behave, and further evidence that much of middle age is driven by the genetic heritage we all share, the 'clock of life'.

And now it appears that the intrinsic, in-built nature of middle-aged social change may have its roots in some fairly basic neural mechanisms. Several studies suggest that people's emotional interactions with others alter during middle age, whether they like it or not. For example, test subjects of all ages find it harder to discern emotion in older adults' faces. Also, as we age, we become less able to discern emotion in the faces of others, and more likely to perceive neutral facial expressions as conveying anger. One report suggests that men actually start to process images of emotional facial expressions in different regions of their brains as middle age progresses. Also, as we get older we become less distressed and angered by interpersonal conflict, and make fewer assumptions about people with whom we argue.

Thus, to some extent our responses to emotion become blunted in middle age – and the volatile emotional reactions of the young start to seem alien. But how can we square this blunting with the idea that middle age is meant to be a time when we strive to form emotional bonds with others? Could it be that, while reinforcing our verbally based emotional links with a select few, we no longer need to be able to interact visually with a large number of strangers and casual acquaintances? In other words, do middle-aged people no longer need such finely tuned emotional radar?

The other kind of conservatism – the political kind – has been reported to increase in middle age in many different countries. So consistent is this finding that many agree there is a fundamental link between age and conservatism, regardless of prevailing ideologies. In the last few years there has also been considerable study into the biological basis of political belief. Some have attempted to show that certain characteristics measured by their psychological tests correlate with political leanings. Neuroscientists have even attempted to dissect the neural components of political belief itself. One study divided it into three elements – individualism, conservatism and radicalism – and identified different regions of the frontal lobes which flare into activity when each political mode is triggered.

However, politics is difficult to study from an evolutionary point of view – partly because we are the only species which 'does' politics, so we have no other creatures with which to draw comparisons. Yet the motives of different political movements may still be amenable to an evolutionary approach. For example, we have already seen that human middle age may involve an inherent tendency to withdraw socially, conserve resources and focus them on our own survival and the success of our kin. Obviously this makes good sense from the point of view of natural selection, but it also sounds pretty similar to the right-wing mindset.

Another aspect of political opinion which makes sense in an evolutionary context is the redistribution of wealth. When we are young we have little, so it makes sense for us to support a system which encourages diversion of resources from the old 'haves' to the young 'have-nots' – a liberal agenda. Conversely, it would be sensible for the middle-aged, the 'haves' – who have most to lose and the least time to spend recouping what is lost – to resist the uncontrolled redistribution of resources, and to support the idea that people should be allowed to use their own property in whatever way they themselves see fit – a conservative political ethos.

We can expect more from this area of research in the coming decade, as brain imaging technology becomes more advanced and more available, and political pundits finally get the chance to do what they have always wanted – to look inside voters' heads to see why they think the way they do. And understanding the way our voting patterns shift in middle age will no doubt be high up on their wish-list.

Middle-aged humans control so much of human society, yet their own identities and tenets are inconstant, fluid things. Why did all those thousands of generations of hunter-gatherer life make us end up this way? It is intriguing to consider what shifting identity, social selectivity and political conservatism might actually mean for someone who lives in a small tribe and owns, effectively, nothing. After all, it is important to remember that the psychological changes we see in today's middle-aged people evolved when life was nothing like it is now.

Yet there must have been a drive acting across those ancient generations to craft the middle-aged personality into its present form. And as a result, in middle age our personality finally gets as near 'completion' as it ever can. But one pressing question remains: does this make us happy?

10. Do middle-aged people really get sadder?

How does one measure happiness? We all have a sense of what it is, we can recognize it in others, we may sometimes even wonder what causes it. We like to believe it is a real entity to which we can contribute, and many of us think it is a major component of a life well lived. We worry that it might melt away as the exuberance of youth gives way to the realities of middle age. We feel it is important, yet we have no obvious way to quantify it – no simple quiz, no blood test, no genetic analysis, not even a coherent verbal definition of what it is, exactly. So how do we measure happiness and what happens to it during middle age?

In early 2008, the results of a survey of happiness throughout the human lifespan escaped from the pages of an academic journal and briefly fluttered through the world's major print and broadcast media. They appeared to show that happiness changes in a systematic way throughout life, and in a way which does middle-aged people no favours. It was claimed that happiness follows a 'U-shaped' curve throughout life, with young people and the elderly perched rejoicingly on the giddy heights of the arms of the U, while middle-aged people languish in the pit of the U's gloomy concavity.

The study involved impressively large numbers of interviewees. The initial survey covered half a million men and women in the United States and Western Europe, but was then extended to Eastern Europe, Latin America and Asia – eventually a total of seventy-two nations. Finally, the study incorporated previously existing data from one million people in the United Kingdom. The results from all these groups appeared surprisingly consistent – that average human happiness reaches its lowest level some time around the fifth decade of life.

Although carried out on a larger scale than ever before, the survey was not revolutionary in how it measured people's happiness – the researchers simply *asked* them. The study was based on questionnaires which allowed subjects to tick boxes labelled 'pretty happy', 'very happy', 'not happy', 'satisfied', 'very satisfied', and so on. This may seem a rather vague form of measurement – and indeed one could certainly question whether 'happy' and 'satisfied' are the same thing – but when studying something as subjective as happiness, perhaps all one can do is ask people's subjective opinions.

Just as important in happiness surveys are the many other questions asked which are apparently unrelated to state of mind – those concerning marital status, children, employment and so on – as the answers provide information about factors which may otherwise confuse the results. For example, let us assume that we are conducting a survey in which middle-aged people report themselves to be less happy than everyone else. Also, in this hypothetical situation, let us assume that people with teenage children tend to be less happy, too. We now have a dilemma because most people with teenage children are middle-aged, so we do not know whether middle-aged people are inherently unhappy because they are middle-aged, or indirectly unhappy because they are likely to have irritating teenage children. Fortunately, this is where statistics comes to our aid. By cleverly comparing all sorts of different people (such as middle-aged people without teenage children, or non-middle-aged people with teenage children), statistical analysis allows us to dissect out the relative importance of all the different factors contributing to happiness. And even after all this statistical jiggery-pokery, the early 2008 study resolutely suggested that middle-aged people are intrinsically less happy than everyone else.

It is easy to think of reasons why young adults should be happier than middle-aged people – beauty, youth, lack of responsibilities, the remoteness of death – but the study's authors came up with some ingenious ideas as to why the elderly, too, might be happier than the middle-aged. One idea was that as people grow old their ambition wanes, their aims become more realistic, and thus they become happier because they fail less often in what they set out to do. A second suggestion was that unhappy people are more likely to die during middle age, thus leaving an 'enriched' population of elderly survivors disproportionately constituted by the happy. Their third theory was that as people pass further from middle age into old age, they become increasingly thankful that they, unlike so many of their contemporaries, are not dead – and this in itself makes them happy. And if I were to subscribe to the U-shaped happiness hypothesis I would add another theory – that the human developmental programme extends even into old age, and tends to make older people essentially happy. However, for such a tendency to have evolved, there would have to be a reason why emotional buoyancy increases the ability of old people to help their offspring succeed.

It is fun to speculate about the U-shaped happiness curve, but it is important to emphasize that however large and carefully analysed the study, and despite the fact that some other studies support its conclusions, there are yet more which do not. For example, one American investigation suggested that happiness gradually increases throughout life, with stressed, anxious young people gradually maturing into the happy, contented elderly.

Also, some psychologists believe that simply asking people if they are 'happy' at one arbitrary moment is not enough. They argue that happiness is not just an indefinable and transient subjective feeling, but the aggregate sum of the positive and negative emotions which people experience (or positive and negative 'affect', as psychologists rather confusingly call it). It is a moot point whether happiness is really the same thing as the summation of all your positive emotions minus the sum of all your negative emotions, but this approach does at least do one thing that scientists like. It takes a big, mysterious phenomenon (happiness) and chops it into small, more readily measurable pieces. In the case of emotions, those small measurable pieces are the numerical scores which volunteers ascribe to each of a list of adjectives on a questionnaire sheet – if they tick low scores for 'disgusted' and 'hostile' and high scores for 'cheerful' and 'excited', then it seems reasonable to conclude that their positive emotions outweigh their negative ones, and thus that they might even be described as 'happy'.

When average emotions of large samples of people are measured in this way, complicated patterns emerge. In middle age, men report more positive emotions than women. In fact, middle-aged men report slightly more positives than younger men, too. Women, on the contrary, experience a slight dip in positive emotions in middle age. Noticeably, in both sexes, positive emotions increase during the transition to old age – beyond the age of sixty. This approach also allows researchers to investigate negative emotions separately from positive ones. Middle-aged men experience fewer negative emotions than women, and their negative emotions decrease a little from their levels in young adulthood. In middle-aged women, negative emotions change little from their youthful levels. And as old age approaches, negative emotions decrease in both sexes.

So what does all this tell us? It seems that middle age accentuates a pre-existing tendency for men to be 'happier' than women (if happiness is what we believe these studies measure). Yet the contrasts between middle age and youth are subtle. I would argue that these subtle differences are what one might expect for two stages of life in which the demands placed upon humans differ in complex ways – and two stages of life which natural selection has honed over millions of years to adapt us to precisely those demands. Once again, middle age emerges as a tightly controlled, distinctive transition, rather than an uncontrolled degeneration. In contrast, once we reach old age we seem to be buoyed up by a much cruder, blanket increase in aggregate happiness.

In addition to all these studies of emotion as an 'aggregate score', there is evidence that what drives our emotions changes throughout our lives. The day-to-day situations in which middle-aged people find themselves have a greater effect on their emotions – daily life often becomes more complicated than before, and contains a more rigid, unremitting routine of increasingly diverse tasks. Although young adults' lives can be demanding, they may find it easier to break away from daily pressures than middle-aged people, and wealth and money is a good example of such pressures. If you temporarily run low on money or food when you are young, it is an inconvenience. Middle-aged people often think back with fondness to their bohemian youth when all they needed was someone to love, somewhere to love them, and a future ahead. In middle age – with its absolute necessity of providing for others and the nagging need to think of the future – it is unsurprising that the demands on our time affect our emotions more.

Yet the consistent shifts in emotion throughout life strongly suggest that this is not about changing external factors. Instead, much of the change is internal. In other words, there is a large developmental component to emotion – the genetic 'clock of life' is changing not only the intensity and importance of different emotions, but also the ways in which environmental influences can affect them. Some researchers suggest that fifty per cent or more of our happiness is genetically determined. This is a striking suggestion, and it leaves surprisingly little room for life experiences to affect our emotions – especially if you consider that our genes may also determine the impact of those experiences at different stages of our life. In fact, everything about our emotional life is beginning to look disquietingly predetermined. And indeed, some have claimed that there are specific psychological processes at work which continually adjust our level of happiness so that it follows its predetermined course throughout life. According to this theory, when our achievements exceed or undershoot our expectations, or our wealth waxes and wanes, we simply respond with a transient emotional reaction which is followed by a speedy resetting of our expectations so that we end up being roughly as happy as we were before. *Plus ça change . . .*

This constant 'recalibration' of emotion corresponds with the everyday observation that some people are inherently optimistic while others are pessimistic, regardless of what life throws at them. Indeed, directly measurable changes in brain function may underlie these tendencies – brain scans suggest that people with a positive approach to life may use their left prefrontal cortex more than their right, whereas the pattern may be reversed in more downbeat people. Of course these studies are quite crude, especially as we do not know whether the differential use of the two sides of the brain causes, or is caused by, positive and negative moods. Yet the idea that differences in personality might be reflected in differences in brain activity patterns does seem plausible.

Investigations into the chemicals released by the working brain show that there may be basic, chemical reasons for emotional change in middle age. Many studies suggest that although middle-aged people's emotions may be sensitive to the sum total of events around them, they actually experience less dramatic emotional responses to *individual* events than younger adults do. This is an intriguing idea, because it implies that middle-aged people are somehow pre-adapted to their super-complex middle-aged lives. We must bear in mind that the variations between the lives of different middle-aged people are probably greater than those between the lives of the young – as regards relationships, achievement, wealth, self-perception, challenges and responsibilities. Thus, to adapt to this inherent variability of middle-aged lives, it is possible that our emotions become more modulated or controlled by the presence of certain chemical neurotransmitters in the brain. After all, middle-aged people's lives are so manifestly (and perhaps unfairly) different from each other's, that if they responded to events like younger people do, their lives would become an unbearable emotional roller-coaster. The last thing you want to do in middle age is revert to experiencing emotions like a teenager does – it would be simply too exhausting, and too disruptive to those towards whom we carry responsibilities.

But rather than signifying a non-specific dulling of emotion in middle age, it is likely that these changes reflect a profound rearrangement of how we think and how we emote. Psychologists enjoy arguing about the relationship between cognition and emotion, but it does seem that the more 'cognitive' regions of our cerebral cortex play an important role in moderating and controlling emotion. For example, we all sometimes experience strong emotional responses to events in the outside world, which we then consciously rein in.

So cognition can control emotion to some extent. This is important because we have already seen how cognition changes in middle age and reaches its life-defining zenith. And indeed, some psychologists believe that it is these cognitive changes which cause middle-age changes in emotion and happiness, too. In other words, middle age is the time when cognition – the feature of humans which has made our species so successful – finally attains its *fully mature* ability to modulate our emotions. We do not become blunted in middle-age: instead the two great realms of human emotion and human thought finally become properly balanced. What could be better than that?

*

To find out more about middle-aged happiness, some psychologists have expanded their research into a related but distinct concept: well-being. Although well-being is as difficult to define as happiness is, it remains a subjectively important entity which differs from happiness in some important ways. First, assessing one's own well-being entails an element of self-appraisal – acquiring an overall sense of our life-success and contentment from how we view our relationships, self-image, autonomy, control of life, purpose and progress. Second, because of this introspective flavour, well-being is a more long-term, considered thing than happiness – it requires us to stand back from the everyday, and review our lives. As a result, well-being is easier to discern from short-term enjoyment, sadness, achievement or failure than happiness is. And third, being social, competitive creatures, part of assessing our own personal well-being inevitably requires us to compare ourselves with other people.

However one defines well-being, its multifaceted, introspective, long-term, comparative nature means that it lends itself to the kind of questionnaires so beloved of psychologists. And the first thing those questionnaires have shown is that people differ in what they consider most important in defining their well-being. For example, some studies suggest that middle-aged people put most value on their relationships, followed by their confidence and self-acceptance. Other studies have even produced ranked lists of what is important to middle-aged people, and one example of such a list is:

- *marriage*
- *wealth*
- *children*
- *health*
- *work*
- *sex*
- *contribution to others*

We will return to 'marriage' shortly, but the relative positions of the other items are intriguing. For example, 'health' appears quite far down the list, but it is important to remember that middle-aged people do not suffer many health problems, so they probably do not think about their health as often as older people do.

'Children' are in a strikingly low position on the list – third seems very disappointing for a part of life which requires so much effort and is, after all, what is important for the perpetuation of our genes. Perhaps middle-aged people put children low on the list because they do not feel that the pressures of raising children and teenagers contribute *positively* to their well-being. Or maybe they see children as a non-negotiable force of nature over which they have no control (I know I do), and instead bias their answers towards factors they feel they can control. Another possibility is that middle-aged people so instinctively incorporate their offspring into their image of themselves, that the thought of considering them an 'external' factor never crosses their mind. Indeed, there is considerable evidence that middle-aged people's assessment of their own well-being is often based largely on their perceptions of their children's successes, failures and happiness, rather than their own. Thus, middle-aged people's well-being may be partly vicarious – as much an assessment of their children as of themselves.

'Wealth' comes high on the list, yet the relationship between money and well-being (and between money and happiness) is a complex one. When people are asked to assess their well-being and their financial situation, yet not asked how linked they believe these two things are, it turns out that income is actually less significant for well-being than other factors, such as mutual support within a family unit, for example. Also, as incomes grow in industrialized countries, people's assessments of their own well-being do become more rosy, but the effect is surprisingly slight – in comparison, increased levels of employment seem to be far more important. According to this argument, we should perhaps be aiming to ensure that everyone is employed rather than prosperous (which is, I suppose, an aspiration not a million miles away from communism). Despite the indistinct links between wealth and well-being, economists have sometimes tried to ascertain monetary equivalents for different life events. When they do this, they come up with surprisingly large numbers – for example, being married has been claimed to contribute as much well-being as an additional £70,000 of annual income, and the death of a partner corresponds to a financial hit of £170,000. (Readers should be aware, however, that I am not claiming that there is a mechanism by which they can actually 'cash in' their marriage for a monetary alternative. Money and well-being are not *that* inter-convertible.)

Perhaps it is not surprising that with all their incisive cognitive skills, middle-aged people rank wealth so highly, since money is the most tangible and measurable item on the list. Yet the perceived importance of wealth remains surprising, considering how little effect wealth has on their well-being. Is the apparent irrelevance of money to well-being yet another example of the power of emotional recalibration? Whenever your income increases, do your lifestyle and emotions simply adapt so that after a few months you no longer notice your pay rise? Alternatively, middle-aged people may have a different relationship with money from younger adults – a relationship in which money is seen as an important means by which they can carry out the unshirkable task of supporting their children, or the altruistic pastime of helping others. To the pragmatic middle-aged mind, wealth seems a far more quantifiable thing than supporting children or helping others, so why not place it high up the well-being list as a talisman of how well you are able to do those other things?

Studies of socio-economic status tell a similar story to those of simple monetary wealth. Although middle-aged people with lower socio-economic status tend to have more health problems, a higher incidence of divorce, parents who die younger and, obviously, less money, their reported levels of well-being are not very different from people higher up the scale. This is a striking finding and rather makes me wonder if we will ever find *anything* tangible or measurable which correlates well with well-being. However, although levels of perceived well-being seem largely unaffected by socio-economic status, the factors which people say contribute to their well-being are certainly affected by it. For example, high-status middle-aged individuals say that purpose, achievement and new learning are important to their sense of well-being, whereas lower status individuals say that being able to adapt themselves to the changing world is more important.

Taken as a whole, however, the outlook for well-being in

middle age is good. Many questionnaire-based studies suggest that self-assessed well-being is greater during middle age than at any other time in adult life. In fact, well-being shows a similar pattern to cognition – a long, broad, reassuring hilltop in middle age. And some studies suggest that middle-age well-being is even more important than that. It appears that middle age is seen by adults of all ages as the defining moment in the story of their lives. Younger adults often base their assessments of current well-being on how well they think they will have done by their forties or fifties – they project themselves forward into their middle-aged future. And conversely, the elderly think of their own middle age when asked to estimate their well-being – casting their mind back to their middle-aged past. It is as if we humans have become fixated on middle age, instinctively taking it as our benchmark of lifelong success and contentment.

This is a crucial part of the power of middle age: it draws our inner, introspective eye, whatever age we are. And if we are satisfied by our middle age – as experienced, anticipated or recalled – it seems we are satisfied by our life.

So what has happened to the U-shaped curve of happiness? Despite its allure, it seems not to correspond well with studies of well-being, nor some other analyses of happiness itself. Could it be that the complexities of human happiness have fooled us?

Indeed, some researchers now claim that the U-shaped curve of happiness is illusory. We have seen that investigating the effects of age on happiness involves meticulous statistical consideration of all sorts of confounding factors, such as children, marriage and employment. And opponents of the U-shaped curve hypothesis have seized on this to show that the apparent U-shape may in fact be a spurious artefact of how these tangled influences were analysed in the original study, rather than an indication of any real saddening effect of middle-age. The most important of these influences is marriage. It seems that marriage, of itself, makes people happier. This also works the other way round – happy single people are more likely to marry, and happy married people are more likely to remain married. The two-way interaction between marriage and happiness is tricky to cope with statistically, but not taking account of this kind of effect can have a surprisingly large impact on statistical results, especially considering that lots of middle-aged people are married and lots of them are happy, too. Indeed, when some of the analyses were repeated with these concerns in mind the U-shape of happiness simply vanished. So we now suspect that the U-shaped curve was never there at all. Maybe middle-aged people do not get sadder.

In this chapter we have explored two overlapping subjective concepts – happiness and well-being. As someone who deals every day with the dry certainties of animal structure and function, tackling these phenomena sometimes seems worryingly vague. However, we all know that happiness and well-being are vital to every one of us, and psychologists have definitely made important inroads into these apparently unscientific phenomena. Indeed, the implications for our 'new story of middle age' are dramatic. It appears that middle-aged people are no sadder than anyone else, and indeed may feel that their well-being is greater than anyone else's. Also, that sense of mid-life well-being seems to be crucial to our sense of life-fulfilment, whatever age we are. And far from being a time of emotional numbness, middle age instead stands out as the phase of life in which the balance between our emotions and our ability to think finally becomes mature.

Having spent the previous chapter emphasizing how our personality remains malleable during middle-age, in this chapter we have seen that people have a striking tendency to retain their optimistic or pessimistic natures throughout life. Of course, each of us continually responds to events around us, but it is clear that our expectations constantly adapt too, and as a result our own personal outlook on the world remains remarkably constant in the face of changing circumstances. Our minds start different, and they remain different. Just as many aspects of human life vary tremendously between individuals, evolution has produced a species in which our personalities also vary a great deal. Indeed, variation in human personalities seems to have been beneficial for its own sake.

Just as humans look and sound very different to each other, natural selection has ensured that we each have a distinctive way of dealing with the world around us, too. And in middle age, that way of thinking about the world – that *personality* – reaches its most mature expression.

11. Is the middle-aged mind fragile?

Defeat maintained a strong association with depression even after controlling for hopelessness [...] whereas the relationship [between hopelessness and depression] was substantially reduced when controlling for defeat. Entrapment and defeat added substantially to the explained variance of depression after controlling for the other social rank variables.

Gilbert and Allan, *Psychological Medicine* (journal), 1998

The human species has controlled, exploited and benefited from the world in a way that no other animal ever has, and the main reason for this is our hugely oversized brain. Although anatomists spent much of the eighteenth and nineteenth centuries looking for the part of the brain which makes us distinctively human – the one crucial evolutionary innovation – they never found it. Instead, sheer size seems to be what makes our brains different from those of all other beings on earth. The human brain is roughly five times larger than one would expect for a mammal of our size (brain size and body size are usually linked in a close mathematical relationship, even in species of greatly differing sizes – mouse brains weigh 0.4 grams and sperm whale brains weigh 8,000 grams, but they both fit the 'standard' mammalian pattern) and we think that this attainment of a 'critical mass' of brain is what allowed us to do all the exceptional things we humans have achieved.

But can it be a coincidence that the one species with a brain so enormous that it can create technology, language and mega-societies, also happens to be the one species in which many individuals spontaneously develop mental illness? I say 'spontaneously' because animals, if sufficiently mistreated, can be induced to exhibit symptoms akin to depression, anxiety and psychosis. Yet many humans – a considerable fraction of the species – develop these conditions without any obvious precipitating cause. One could argue that mental illness is the result of the human brain growing towards its absolute limits of size and complexity. We are approaching the point at which our brains are so big that they are becoming unstable.

And many people, especially the young, believe that middle age is a time when the distinctively human burden of mental illness is at its greatest. Middle age is often seen as an inherently grim time, a phase of life when we are most vulnerable to the three human scourges of depression, anxiety and psychosis. Yet in this chapter I will consider whether this is indeed the case. I will focus on depression, with occasional references to the other two illnesses, as I think it is the mental illness most often associated with middle age.

Clinical depression in middle age is certainly important. Any of us may experience intense sadness – 'reactive depression' – in response to adverse events in life, but 'clinical depression' comes more from within the individual. Although the distinction between 'reactive' and 'clinical' depression is more blurred than psychiatrists used to believe, the latter characteristically involves protracted periods of unremitting sadness, with self-defeating cycles of guilt, self-blame and feelings of worthlessness. Unlike reactive depression, clinical depression makes it almost impossible to appreciate the brighter side of life, and sufferers often experience distinctive physical symptoms as part of their illness, such as early-morning insomnia, headaches and digestive problems

The all-consuming nature of clinical depression means that it can be devastating in middle age. Sufferers often function poorly in social situations, retire early into late-life poverty, and are more prone to chronic pain, diabetes, heart disease and obesity. The mental effects of depression are not only restricted to mood, but also linked to reduced memory and cognitive performance, as well as unusual patterns of brain activity on brain scans which can persist even after episodes of depression have ended. Fortunately, however, surveys of depression in the population suggest that, contrary to popular expectation, we are actually less likely to suffer from depression in middle age than at any other time in our lives. The chances of a woman in the developed world experiencing an episode of clinical depression in any given year are approximately twenty per cent before middle age. Between early and mid-middle age this incidence decreases to sixteen per cent, and by the end of middle age it is perhaps ten per cent. Men seem to be less prone to depression at any age – I have already mentioned that their emotions are more positive in middle age, that they tend to overconfidently estimate their own physical health, and that they believe themselves to be unrealistically far from death. Their annual incidence of clinical depression is approximately twelve per cent until mid-middle age, and thereafter it declines to seven per cent and perhaps as low as three per cent in old age. It is important to emphasize that these figures still indicate that large numbers of people are affected by depression at all stages of life, but the trends in middle age do undeniably seem to be going in the right direction. (In fact, anxiety disorders exhibit a similar decline over the course of middle age, and schizophrenia, also, is less likely to start in middle age than during the teenage or young adult years.)

Psychologists have thought long and hard about the apparent decline in depression in middle age, because they want to be sure it is real. For example, it has been suggested that middle-aged people might be more likely than younger people to forget depressive episodes when they take part in mental health surveys. However, as well as lower rates of 'remembered' depressive episodes, they also consistently report lower rates of 'current' depression, so it is unlikely that middle-aged people are simply failing to recall earlier periods of depression. A second possibility is that as people get older, they may become more confused and thus not think to report episodes of depression, past or present, but while this may be a problem when gathering information from the elderly, significant confusion is not a common feature of middle age – even during bouts of clinical depression. A third possibility is that middle-aged people might be less willing to admit to depression, or less willing to seek treatment. But this too is refuted by the data – in fact, some psychologists claim that middle-aged people are more inclined to report their negative moods than younger adults. I would add a further, fourth reason why middle-age depression might be under-reported, at least in comparison to teenagers or the elderly. Teenagers are regimented in school, and old people tend to be under relatively frequent surveillance by their doctor, so one might expect that depression is less likely to go undocumented in those age groups than in middle-aged people. Yet this still does not explain the decline in rates of depression between early adulthood and middle age, since young adults are just as unregimented and under-observed as the middle-aged.

So if depression really does decline by the time we reach middle age, which middle-aged people still suffer from it? There have been many studies which have attempted to answer this question but as we will see, discerning what causes depression, what is caused by depression, and what merely happens to be caused by the same things which cause depression, can be a confusing process.

For example, married people are less likely to report symptoms of depression. This is probably because they experience less depression than unmarried people, rather than because married people prefer to keep their depression secret. Yet, it still does not necessarily mean that marriage is inherently 'protective' against depression – one could just as well argue that people prone to depression are less likely to marry, or stay married. (The same could be said of schizophrenia, which is also less common in married people.) However, one finding which suggests that marriage might indeed prevent depression is that being widowed seems to have the opposite effect – it correlates strongly with depression, and the effect of losing a spouse appears to be more dramatic in men. And intriguingly, the correlation between marriage and 'non-depression' is also more marked for men than for women. In women, it is not the act of becoming married which seems important, but rather the perceived state, or success, of the marriage.

The effect of children on depression in middle age is unclear. Some studies have shown no link between having children and the incidence of depression, whereas others suggest that parents, especially mothers, are more likely to suffer from depression than non-parents. The correlations are so weak, however, that this may not be a real effect at all. For example, pre-existing depression is known to be linked to a greater likelihood of having children relatively early in life. Thus, children may not be causing depression at all – rather, depression may be 'causing' children. Indeed, there are probably good evolutionary reasons why young humans faced with adversity respond by having children – early-life adversity is a sign that life may be harsh and short, so the best response may be to start breeding as soon as possible.

Although I was rather dismissive of Sigmund Freud earlier, there is little doubt that people's formative experiences can influence their chances of suffering from depression in middle age. In one UK study, forty-three-year olds were asked to state whether they felt that their parents had treated them in a way which they would describe as 'caring', 'controlling' or 'non-engaged'. When the results were subsequently compared with the same cohort's mental health at fifty-two, a recollection of 'caring' parents in childhood was significantly linked to good mental health, as compared with the 'controlling' and 'non-engaged' groups. Other studies have shown that depression is also twice as common in middle-aged people who were sexually or physically abused as children. In addition, the length of early-life education appears to have a strong preventive effect on depression in middle age. However, all these apparent links must be treated with caution – because what is cause and what is effect is often unclear. For example, are depressed people less likely to describe their parents as 'caring'? Is it easier to continue in education if you do not suffer from mental illness? What burden of prior mental illness do previously abused people already carry with them into middle age?

Paid employment correlates so strongly with middle-age

mental health that it is tempting to accept that it protects against mental illness, although it is disconcerting to note that the incidence of depression in homemakers is reported to be similar to that of the unemployed. Also, at any given age, the apparently beneficial effect of paid employment on the mental health of men is greater than that for women, and this leads to some intriguing differences between the sexes. The proportion of men who are employed is usually highest in the third decade of life, and then gradually decreases throughout life – which means that middle-aged men have lower rates of depression *despite* the fact that fewer of them are working. In contrast, because many younger women remain at home to care for children, the peak age of employment in women is around the age of forty, when many enter the job market once their children have grown up. Thus, increased employment in middle age may be actually *contributing* to falling levels of depression in women.

Some psychologists believe that the most important influence on depression in middle age is people's position in their social hierarchy. Indeed, they suspect that many of the proposed links between marriage, parental care, education and employment, are simply manifestations of an overarching effect of social position. In common with many social animals – red deer, chimpanzees and mole rats, for example – groups of humans naturally arrange themselves into social hierarchies. This process begins in earnest in the early teenage years and it is likely that these hierarchies, and the criteria on which they are based – beauty, health, wealth, employment and intelligence – do not change very much by middle age. From adolescence onwards, human populations spontaneously establish two separate pecking-orders – one for men and one for women – and although they may ascend and descend in those hierarchies, they cannot escape them. Middle-aged women, especially, face discomforting challenges to their social status, as studies show that beauty is a far more important determinant of social dominance in women than in men. Thus, at a time when many men are enjoying social elevation due to increasing wealth and power, women may feel their wrinkles are dragging them in the opposite direction. We know that people find being low in social hierarchies uncomfortable, and indeed, animals in equivalent positions also exhibit obvious signs of stress, such as abnormal behaviour, counterproductive social interactions and poor health. So the idea has grown up that mental disease is an aberrant human manifestation of a problem faced by all low-status social animals. And the quotation at the start of this chapter shows how concepts such as hopelessness, defeat, entrapment and general social failure have now been incorporated into the language of mental illness.

Other studies point to even more mechanistic, physiological, 'body-as-machine' origins of depression in middle age. For example, when presented with stressful experimental situations, most of us react by increasing our blood pressure and heart rate, and secreting a hormone called cortisol from our adrenal glands. However, in people with clinical depression these stress responses are reduced – and in fact the degree to which they are reduced corresponds closely with the severity of their depressive symptoms. Even when researchers statistically exclude the effects of medication, recreational drugs, health and socio-economic status, this link with the hormonal stress response still remains. The findings are supported by the results of another study, in which middle-aged women who reported feeling more stressed showed lower increases in cortisol when they woke up in the morning, and they also had lower levels of cortisol throughout the day.

These links between middle-age psychology and middle-age physiology are intriguing, and the stakes are high, as not only can middle-age depression be devastating in itself, but also it is linked with other serious problems. In middle-aged women there is a strong connection between depression and low levels of physical activity and excessive caloric intake, and thus, unsurprisingly, obesity. Also, depression is commoner in middle-aged individuals who consider themselves – rightly or wrongly – to be overweight. In men, there are strong links between depression and cardiovascular disease, although in this case it seems that either depression may become evident before the cardiovascular disease does, or vice versa. Later on we will look at how sexual change, too, can impinge on our mental development. In all these ways, body may affect mind, mind may affect body, and the 'clock of life' may hide within it surprises which can affect both body *and* mind in middle age.

Although pre-existing causes of depression such as genetics, health and life-situation are undoubtedly important, the other critical components of mid-life mental health are the factors which actually induce stress. Psychologists used to see middle age as a sequence of repeated bouts of psychological pressure and crisis resulting from the continuing process of personal, psychological change. However, more recently the experts have started to agree with the commonly expressed view of middle-aged people that it is the small, everyday, low-grade irritations which can drag them down into inert, malignant sadness. Marital problems or even managing a successful marriage come high on the list of stress-inducers, as do health problems, perhaps because middle-aged people do not expect to get ill very much. Children, too, can be a trigger – after all, they are often demanding, stroppy teenagers, or are leaving home. Increased power in the workplace brings greater responsibility and pressure, and middle age is a time when women are often entering or re-entering the workforce during middle age, and some men are already leaving it, voluntarily or not. Middle-aged people's parents are often becoming unwell. And despite the fact that immediate financial stress declines during middle age, long-term financial planning becomes a pressing concern. In general, it is women's lives which are changing the most, so it comes as no surprise when studies show that middle-aged women's well-being is very dependent on how well they feel they are coping with changing family roles, career and relationships.

Stress is not just about long-term, overarching, 'big' pressures, but also the small-scale, everyday niggling irritations of life at home and at work, and some studies suggest that the nature of these irritants changes in middle age. First of all, middle-aged people report fewer 'stressors' per day than young adults, despite the fact that they may appear to face a wider range of tasks and responsibilities. Also, in middle age, women report more stressors per day than men – and the most common short-term stressor for both sexes is their own children (sound familiar?). Yet middle-aged people report that fewer of their day-to-day stressors are 'frustrating' than younger adults do. In other words, things which cause stress in middle age tend to be things about which middle-aged people feel they can do something – they are things they believe they can control and, as we saw earlier, this can be a very comforting thing.

*

So things are looking good for mental health in middle age. People are probably happier, feel more in control and suffer fewer stressors, especially 'frustrating' ones. Many psychologists think that middle-aged people develop a new, effective personality in which emotion and cognition come into balance. All in all, perhaps it is unsurprising that the incidence of mental illness really does seem to decrease in middle age.

Even those psychologists who are pessimistic about middle-aged people's emotions accept that middle age does not seem to lead to increased levels of depression. Indeed, they point out that falling levels of depression imply that there is something inherently *resilient* about middle-aged people – they can experience sadness and stress, yet they can somehow prevent this progressing to pathological mental illness.

Recently, several possible causes of this middle-age mental resilience have been proposed. The first is that middle-aged people experience stress less frequently than younger adults – and we have just seen that there may be some truth in this. A second possibility is that people who suffer from clinical depression are more likely to die before they reach middle age, thus leaving a surviving middle-aged population inherently less likely to become depressed. A third theory is that middle-aged people are protected from mental illness by a generally reduced tendency to react to stress. This might involve the subtle emotional detachment and refocusing which occurs in middle age, or more simply the fact that middle-aged people have more resources from which to draw mental support – wealth, family, certainty, and so on.

A final possibility is that the wonderful balance between thinking and emotion which emerges in middle age, and the accumulation of several decades of life experience, create a personality which is inherently stable. It seems that the middle-aged mind avoids and manages stress with an effectiveness unmatched by people in any other phase of life. Indeed, middle age represents a moment of final, maximal, psychological development – a time when our ability to cope with the bad aspects of life predominates over our tendency to ignore them or make illogical or even delusional reactions to them.

So, the natural mental strategy of a middle-aged person is to actively manage their own emotional responses. Middle-aged people finally learn how to divert their emotions away from self-defeating cycles of self-criticism and channel them towards dealing with life in a purposeful, focused and effective way. And in this context, middle age really is the pinnacle of that huge, all-conquering human brain. Middle-aged people are not standing somewhere in the middle of a vaguely rising slope of improving mental resilience. Instead they are at the summit of a resilience-mountain. Young adults are more likely to experience mental illness than the middle-aged, and the elderly are more likely to experience relapses after episodes of mental illness. Middle-aged people are, quite simply, more able to cope. These provisioners of the tribe, these compulsive conveyors of culture, are so important that they have evolved to be an island of stability in the maelstrom of human life.

12. So . . . what is the secret of a flourishing middle-aged mind?

We have spent the middle third of this book focusing on a single organ of the middle-aged body. This may seem disproportionate, but I would argue there are good reasons for giving such emphasis to the brain.

As I have already argued, the brain is the one thing which sets our species apart from the beasts – the thing which makes humans essentially different. Although there are other things which make humans unusual – walking on two legs; manipulating, tool-fashioning hands; exotic social and sexual behaviour – every one of these is bound up with the evolution of that capacious mind. Indeed, in the last part of this book we will see just how exceptionally dominant the human brain has become over the rest of our biology.

The other reason for my brain-bias is that I thought the brain would be especially interesting to you, dear reader. After all, your brain is *you* in a way that other organs simply are not. It is where your self and personality reside and it is whence you perceive the world outside. It is the place my words are popping into right now. And when you reach middle age/now you are in middle age/when you once were middle-aged (delete as applicable), middle age will be/is/was a phenomenon characterized by a constant interaction between your changing brain, your changing body and the changing world outside.

In addition to this, middle age is the stage of life in which the brain becomes, if anything, even more important. As physical and reproductive attributes wane, it is the brain which powers us forward into a bright and productive future. Indeed, middle age is a time when many people believe that their brain is their main remaining 'saleable' resource – the best of what they have to offer.

Before we leave the brain behind, I want to take a different perspective on what we have learnt about the middle-aged human brain in recent years. Up to now we have picked our way through cognition, time perception, personality, happiness and mental resilience, but I have tended to take the usual scientific approach of 'looking at averages' – whether average middle-aged people differ from average young adults, or whether, on average, mid-life men differ from mid-life women in some way or other. There is nothing wrong with this approach, but none of us ever lives an average life, and certainly none of us ever lives an average middle age. There is variation in everything that happens in the middle-aged brain, so some of us get lucky and some of us do not. We all know that life is unfair, but in middle age this unfairness becomes especially dramatic. Quite simply, people's hopes, abilities and achievements may either thrive or founder in middle age, and often not because of anything that could be considered their own fault.

So, enough of averages: what about individuals? What *does* determine how individual brains and minds change in middle age, and what does this mean each of us can expect in our fifth and sixth decades? How does each one of us adapt to the evolutionarily imposed mantle of middle-aged super-provider and culture-conveyer?

Inequality is rife in middle age. Some people achieve great things in life while others do not, and middle age is the time when we must come to terms with this fact. After all, there are ever-shrinking opportunities to change matters in the decades still left. I have been very positive about many aspects of the human mind so far, but while minds may improve during middle age, minds may deteriorate, too. Furthermore, both improvement and deterioration seem to be self-perpetuating in middle age – exaggerating rather than narrowing the differences between individuals.

Inequality is, by its nature, unpredictable. However, studies suggest that there are common factors which help to maintain a flourishing middle-aged mind. Several of them we can control, but many we cannot. Here are some of them:

1. *Being high in the pecking order.* Socio-economic status has a dominant positive effect on general physical health in middle age and beyond, and this seems to have a strong knock-on effect on mental performance and stability. It is known, for example, that major physical illnesses result in measurable diminution in cognitive abilities. Similarly, recurrence of depressive illness following a single episode is more likely in people who have concomitant physical illnesses, or who have lower socio-economic status. Because of these effects, many developed countries are a demographic patchwork of adjacent regions of high and low socio-economic attainment, high and low physical health, and high and low middle-aged mental well-being.

2. *Having a long education.* Socio-economic status has a strong link to education. Researchers like using education in

their studies because it is relatively easy to measure – simply asking people at what age they left full-time education provides a clear, simple and useful number to use in statistical calculations. For example, level of education has clear links with physical health, positive emotions and maintenance of cognitive abilities in later life. People argue about cause and effect, but the analyses suggest that better educated people make life-choices which render them more physically healthy in middle age, which in turn promotes mid-life mental function. (Although this does not rule out the probability that people who stay in education for longer might have been destined to be more cognitively able anyway; nor does it exclude the possibility that the process of education itself may directly promote beneficial mental processes in middle age.) Another phenomenon, which might loosely be considered part of someone's education, is bilingualism. Being bilingual has been shown to confer clear cognitive advantages – especially, it seems, in the ability to pay attention. And remarkably, this beneficial effect persists well into middle age.

3. *Being employed.* Although it is difficult to extricate the effects of employment from those of socio-economic status and education, there is some evidence that more highly paid, white-collar jobs tend to promote continuing cognitive flexibility. I suppose this could be rephrased as mentally challenging jobs stopping middle-aged people getting into a rut. Research also shows that the effects of working in middle-age can be especially beneficial for women, many of whom are 'emerging' from a period of childcare to enter the labour market. However, studies suggest that work, like many things, should be taken in moderation – there is also clear evidence that working long hours (fifty-five hours instead of forty hours per week) can have adverse effects on middle-

aged people's performance in reasoning and verbal tests.

4. *Being around people who talk you up.* Social perceptions are very important in how middle-aged people view their own success at work. Although life rarely ends during middle age, careers often do, and the workplace is often a hotbed of ageism. Ageism has real, measurable effects on how people function – some remarkable studies show that older adults perform better on cognitive tests if they have simply been exposed to positive images of older people's abilities immediately before the test.

5. *Being female.* Sex is a major factor which influences the cognitive journey taken by the middle-aged mind. While middle-age changes in mood and mental health may not be as beneficial for women as they are for men, changes in cognition may benefit women more. It appears that women's middle-age improvement in cognitive ability is slightly greater than that of men, and that the summit of women's 'cognitive hill' comes slightly later. There are some aspects of cognition which receive less of a boost in middle-aged women – speed of perception, for example – but these may reflect distinctive evolutionary adaptations of the two sexes. It is clear from simple observation that the female human body is not adapted for hunting and foraging over large geographical ranges, so it might come as no surprise if the female brain turns out to be less adapted for these activities, too. And these differences between the habitual activities of ancient humans probably also explain the gender disparity in cognitive skills – just as at every other stage of life, middle-aged women perform better in verbal, semantic tasks, and men perform better in visual, spatial tasks.

6. *Keeping healthy.* Although you cannot do anything about your sex, and may not feel you can do much about

your socio-economic, employment or educational status either, there is one aspect of life affecting the middle-aged mind which you *can* voluntarily change, and that is your physical health. Obviously, health in middle age has large 'uncontrollable' genetic and social components, but it is at least amenable to deliberate change in middle age. And some of the links between physical health and mental ability are surprisingly specific. Obesity, for example, is correlated with reduced 'semantic' memory – general, everyday, impersonal memory of the world around us. Conversely, physical activity is linked to higher cognitive functioning in middle age, as are leisure activities which demand social or cognitive effort. Cardiovascular health also corresponds with high cognitive functioning. For example, cardiovascular changes which have not yet caused obvious symptoms, including high blood pressure, can still adversely affect mental function. Circulating levels of cholesterol are associated with cognitive deterioration in middle age as one might expect, but this link disappears as we enter old age – perhaps because a sufficient number of people with cardiovascular disease die in middle age so that the elderly population is already purged of the tendency for vascular disease to diminish brain function.

7. *Drinking alcohol.* Alcohol consumption has unexpected effects on mental processes, some of which the quaffing classes may find reassuring. Although alcohol has obvious links to obesity and ill-health, and can serve as a chemical prop for people with mental and physical health problems, it may not be a complete villain – there is, for example, tentative evidence that it may slightly reduce the risk of cardiovascular disease in some individuals. Some studies have suggested that drinking at least one alcoholic drink each week may improve middle-aged people's performance in a

variety of cognitive tests, and enhance cognitive flexibility and even cognitive speed (not *immediately* after consuming the alcohol, of course). The possible beneficial effect of alcohol seems to be more pronounced in women, and remarkably, it is still present in people who drink quite large amounts – more than the UK government recommends. These findings suggest that we must adopt a sophisticated approach to dealing with middle-aged intake of alcohol – that plentiful, natural drug to which we have probably been most exposed during our evolutionary past. Could our occasional ancient slurping of fermenting fruit have left a beneficial legacy in our genes?

8. Having the right parents. Finally, if there is one aspect of human mental ability which people shy away from, it is the idea that cognitive ability is an inherited trait. Many people find slightly disturbing the suggestion that intelligence – the characteristic which has done most to make us successful – is partially controlled by the genes we inherit. Because of this, discussions of this topic are often emotionally charged in a way rarely seen in other areas of science. Yet if cognitive ability is so important to humans, it would be bizarre if it were not to some extent heritable – after all, the alternative would be for it to be the product of an ill-defined, hit-and-miss interaction between a developing person and their environment. Cognitive ability is built into the human species, so there is no reason why it should not to some extent be built into every one of us, too – imposed like so many other things by the ticking genetic-developmental 'clock of life'. Indeed, some researchers have provided evidence that cognitive ability in middle age is actually one of the most heritable traits we humans possess. We can add mid-life cognitive ability to the list of human characteristics unequally and unfairly dealt out to different individuals at the whim of genetic inheritance, a list which already includes beauty, health and rate of ageing.

The middle-aged mind fits well with our triad of middle age: it is distinctive and unique, and some of the changes it undergoes can be abrupt. Yet whatever middle-aged mind we each end up with, there is a time when we must stand back and explore what all this neural, psychic and emotional change means for us as an individual, and what it means for human society. What does it feel like to inhabit a middle-aged mind? And why does that make middle-aged people good at what they do – how do they provide and communicate so effectively, and why do they seem to run the world? In addressing these questions, we will also see how middle-aged behaviour has its origins in the everyday lives of ancient humans.

Throughout this exploration it is important to bear in mind that the middle-aged mind does not exist in a vacuum. The world is also full of people who are not middle-aged, and this is where the perfect complementarity of young and middle-aged adults starts to becomes evident. We will see that once we break down some of the clichés of middle age and take a more evolutionary approach, the role of middle age in the onward striving of the human race becomes clear.

A good example of just such a cliché is that middle age makes people dogmatic and irascible. Many earlier studies emphasized that middle-aged people are more inclined to maintain their own opinions in the face of disagreement, and also prefer to stick to established routines and procedures. It was also claimed that this rigidity, along with a tendency to misjudge the attitudes of others (especially the young, who are frequently exasperated by this apparent obstinacy), causes many of the conflicts which take place in human societies.

However, we have already seen that middle-aged people often do have superior cognitive abilities, and that they may be less emotionally reactive and less sensitive to the emotions of others. So, if middle-aged people are indeed dogmatic, I would suggest that this is for a good evolutionary reason. Middle-aged people control the day-to-day activities of most human societies – even those societies apparently run by cliques of 'elders' – so they have a vested interest in being decisive and consistent. Also, if a major function of middle-aged people is to convey culture (in its widest sense) to younger generations, then at some stage they must 'fix' their concept of what that culture actually consists of. Middle-aged people must transmit a coherent and unchanging representation of what they see as human culture – so skills, technology, attitudes, beliefs and art must at some stage be codified in rigid form to be efficiently bequeathed to subsequent generations. However, because of this tendency to 'fix' perceptions of culture in middle age, middle-aged people can sometimes appear cranky and inflexible to the young. Indeed, certain strange behaviours seem to establish themselves as normal in the minds of middle-aged people, and this process is further enhanced if male-female partners accept, mimic and reinforce each other's weird foibles – a process rather wonderfully named 'behaviour contagion'.

That said, some studies suggest that middle-aged people are not actually as inflexible as was once claimed – or at least that they are inflexible in subtle ways. Indeed, there is some evidence that throughout the course of middle age, people become more risk-averse and thus less keen to make independent decisions. Hence, if they truly are becoming dogmatic, it is because they develop a tendency to dogmatically seek consensus before decisions are made – hardly the autocratic dogmatism of which middle-aged people are often accused.

As well as changing the way in which it asserts itself and makes decisions, the middle-aged brain is also adept at turning its attention inwards. Developing an ability to self-criticize is a generally beneficial, occasionally damaging, but nonetheless essential part of growing up and becoming fully human. In fact, the ability to analyse one's own abilities and weaknesses, successes and failures, and to dissect one's own thought processes, is a central part of human life. If an adult human fails at a task, then their immediate reaction is not to stubbornly try again, but instead to think long and hard about their own performance and to work out where they went wrong and how they can succeed in future.

In middle age, this self-analysis takes on a new flavour. Middle-aged people think a great deal about the perceived loss of ability which comes beyond the age of forty (and some even feel driven to write books about it). Yet there is something about middle age which invites such self-examination. For example, it is the most challenging life-phase as far as work and achievement are concerned – middle-aged people often feel that many productive years are still required of them (unlike old people) but fear that their physical and mental abilities are deteriorating (unlike young adults). Also, the tacit employer-employee pact of permanent, supportive employment and generous old-age provision which predominated in the decades after the Second World War has evaporated, leaving today's middle-aged people in an unexpectedly precarious position. All this self-doubt is exacerbated by the fact that middle-aged people often retain a distortedly rosy view of their abilities and vivacity during their youth. Indeed, studies show that while middle-aged people may clearly remember what their life was like five years ago, their recall of life twenty years ago is often poor, meaning that middle-aged people may be comparing themselves to a glorious past that never was.

Of course, the converse of all this self-criticism and worry is that middle-aged people are able to identify their strengths and focus on the positive, wherever they find it. And socio-economic status and employment type seem to be very important in determining how people perceive middle age itself. While blue-collar workers tend to say middle age starts at forty, white-collar workers tend to choose a figure nearer fifty, which suggests that middle age is not only a set of distinctive biological changes which occur at slightly different times in different people, but also seems to be a state of mind.

Self-analysis is also the key to middle-aged people's successful interactions with the young – and it is hard to emphasize sufficiently just how important this interaction is for our species' continuing success. Human society and culture are created by constant interplay between two very different types of people – teenagers/young adults and the middle-aged. Throughout human history, young people have been good at novelty, creativity and cultural change, and middle-aged people have been good at analysis, planning, organization and cultural continuity. This specialization in our ways of thinking is not socially imposed – it is the unavoidable result of structural and functional changes which occur in our brains at different stages of life. You cannot avoid having a young brain when you are young, just as much as you cannot prevent your brain getting middle-aged. The continuing development of the human brain throughout adult life is not optional – it is indelibly stamped into your genes at conception. And because of this, to this day there persists an uneasy, unending tussle for supremacy between the two great human age-cohorts – the young trying to change things and the middle-aged trying to perpetuate what has worked best in the past.

To participate in this tussle, middle-aged people must be able to keep up with, and often keep ahead of, younger, brighter, quicker people – but how? Rather than 'experience', I suspect that middle-aged people outperform young adults by using something we can call 'perspective': studies suggest that middle-aged people are especially good at 'seeing the wood for the trees'. Experiments show that they can hold larger batches of information in their heads, and can also 'take a step back' and view this information globally and in context, rather than being confused by details. For example, evidence from middle-aged typists suggests that they hold longer strings of text in their heads to compensate for the fact that they cannot type as fast as they used to; research on middle-aged engineers reveals that they are more able to sift and select from novel information to simplify problems and avoid confusion; and middle-aged salesmen have been shown to spontaneously develop entirely new ways of selling, tailored to optimize their success as they grow older.

All these adaptations reduce the amount of energy required to perform mental tasks, and as we saw earlier, middle-aged humans are very energy-efficient. Middle-aged people are also better at delegating tasks and responsibilities, probably because their global perspective makes it easier to guide others, especially the young (management consultants like studying this sort of thing). They find it easy to explain not only what needs to be done, but also why particular tasks are important for the success of the shared endeavour. And ancient human life was all about shared endeavour – the tribe cooperating to achieve some mutual goal, knowing very well that middle-aged and young adults had different skills to contribute. Another benefit of new-found mid-life perspective is that it makes it easy for middle-aged people to set clear priorities and goals, too – an ability which in the past we might have called 'wisdom'.

So much of middle-aged behaviour – what it is like to be, or to interact with, a middle-aged person – harks back to our evolutionary past. For example, we believe that a major role of middle-aged men was to leave the tribe temporarily and diligently acquire resources. Perhaps this explains why middle-aged men enjoy leaving the family dwelling at weekends and dealing with their favourite concrete and abstract things, sometimes alone, and sometimes together in the matey and relaxing all-male atmosphere of the riverbank, the sports ground, the garage, or most of all, the cherished shed.

One might worry that this middle-aged cognitive change and inter-generational tussle could thwart human progress – especially in a modern world in which the rate of cultural change is now so fast that disagreements between young and middle-aged adults are enormously amplified. Despite this, I propose that this inter-generational conflict is beneficial – that it leads to a continual productive, creative tension in human life. Without the young, little would ever change, yet without the middle-aged, there would be no cultural memory and human life would become chaotic. Useful human life must be a balance between change and continuity. After all, what would the young do with their time if they did not have an older generation against which to rebel?

PART III

OLDER AND BOLDER

Romance, Love, Sex, Babies and Life After Forty

Then Abraham fell upon his face, and laughed, and said in his heart, Shall a child be born unto him that is an hundred years old? And shall Sarah, that is ninety years old, bear?

Genesis 17:17

13. The end of sex? (An introduction.)

The sheer weirdness of humans is never more apparent than in the morass of love and sex. Trying to understand middle-age sex is an even more daunting task because so many of our species' unusual features converge and collide in it. It is neither so beautiful that we revel in it, nor so revolting that we shy away from it, so usually we just joke about it. Middle-aged love and sex remains a tangled knot of confusing, conflicting forces which dominates the middle decades of our lives.

Untangling this knot will take us our six last chapters, but the central paradox is simple. Humans have far fewer babies after the age of forty than they do in the two preceding decades, and this has always been the case. Yet we often live to twice that age, and as we saw earlier, this too has probably been true for most of human history. So where does this leave sex and relationships beyond forty? What is the point of sex when there are no more babies to be had? As we will see, romance and sex are eventually stripped of their procreative overtones in middle age, and the very essence of the human sexual condition is exposed. Once sex loses its baby-making function, all that remains is its humanness.

I will retain my evolutionary-zoological approach as we

tackle these questions, because characterizing the utter strangeness of the human species allows us to best understand what life throws at us. Today's middle-aged relationships are direct consequences of millions of years of natural selection, which cast some of us aside and allowed others to prosper. We are descended from the ancient humans who survived, and they survived because they acquired some pretty strong urges. One striking result of all this brutal evolution is that men and women have become very different from each other – exaggeratedly so, in fact. We seek different things, need different things, and approach life in completely different ways. Sometimes we are in conflict.

Over the next few chapters we will describe what happens to the four fundamental reproductive forces – the sexual act, female sexuality, male sexuality, and the ability to make babies – as the potential for procreation wanes during middle age. Finally, we will turn our attention to how these upheavals affect families, and the romantic relationship itself.

So how often do middle-aged people have sex? The simple answer is that they have more sex than younger people might like to think they do.

Early research into sexuality which took place in the United States in the 1940s and 1950s suggested that there is a precipitous decline in heterosexual couples' sex lives as they progress through young adulthood and middle age. Intriguingly, of all the possible sexual behaviours available to a couple, it was vaginal intercourse – the practice most likely to generate offspring – which seemed to decline at the greatest rate. Of course, viewed from a different point of view, this suggests that sexual behaviours which do *not* make babies are selectively *preserved* as humans age. Apparently, something else supersedes fertility as the main reason for sex – and we will return to this idea later.

Some more recent research has been aimed specifically at middle age. One US study suggested that among people 'leaving' middle age – aged between fifty-seven and sixty-five – seventy-three per cent report themselves as being sexually active, but this figure slumps to twenty-six per cent two decades later. Bearing in mind that research also suggests that young adults are not always as sexually active as one might assume, then sexual activity in middle age seems comparatively buoyant.

However, the fact that sex is happening does not necessarily mean that it is happening as often as it did before. In retrospective studies of long-term marriages, the frequency of sex has been claimed to decline dramatically in the first year – by as much as a half. At this early stage, the frequency of sex seems to decline more rapidly in couples who did not have pre-marital sex. The speed of decline may be even more dramatic if the date of the commencement of sexual relations is used as the starting point, rather than the date of marriage. And as if that were not enough, this initial decline is then thought to be followed by a *second* halving in sexual frequency, although this takes place at a far slower pace over the course of the next twenty years.

At best, these are average figures and one suspects that even if they are true, individual couples often deviate from them considerably. Indeed, there are factors other than the length of a sexual relationship which influence sexual frequency as we get older. For example, healthy middle-aged couples are fifty to eighty per cent more likely to report interest in sex that those in poorer health, and the elderly often ascribe their low frequency of sex to deteriorating health. For women, the menopause has been claimed to be an important factor, yet the impact of the menopause on sexuality remains controversial – the effects of chronological age and the menopause can be surprisingly difficult to disentangle, and some claim that the menopause has no direct effect at all. Another possible factor is parenthood: having a baby is often claimed to cause a spectacular decline in the frequency of sex, and the data certainly suggest that this is a real effect, and one which is permanent rather than transient. Other intriguing findings include those which suggest that mixed-race couples have more sex than same-race couples, and that couples who argue more also have more sex. Also, while a stated intention to make a baby does increase the frequency of sex, its effect is surprisingly weak, even in younger adults. It is almost as if humans are somehow separating the desire to have sex from the urge to have children, which from the evolutionary point of view is a dangerous path to tread. Natural selection does not usually smile upon those with a take-it-or-leave-it attitude to procreation.

And how much sex people actually want is an important issue. If middle-aged people have half as much sex as younger adults, then maybe this simply reflects a lower level of desire, although here the data are confusing. One aspect of sexual behaviour which correlates well with women's relationship satisfaction is the frequency of vaginal sex – this is the one activity with strong links to women's reported sense of intimacy, trust and love. However, when explicitly asked their opinion, more than a quarter of women in an Australian study aged between forty and forty-nine claimed they were indifferent to how often they had sex. Also, while studies suggest that men are more likely to lament the declining frequency of sex, women seem more prone to worry about the 'quality' of sex. Ability to reach orgasm is also important for women, but whether this is for its own sake, or because it often represents the 'successful' culmination of that all-important vaginal sex is unclear. So our conclusions about sexual frequency remain uncertain, although it seems clear that the sex people *want* is just as relevant as the sex they *get*. Both sexes appear to be happier if the sexual frequency they attain matches their expectations.

Whatever the surveys show, it should be emphasized that almost all studies of the frequency of sex are carried out on people living in the modern, post-agricultural, post-industrial world. This is hardly surprising, but we have already seen that human life was very different during most of human history, and that we are the evolutionary products of that earlier age, now born into a new, 'unnatural' environment. Because of this, it is entirely possible that many of the trends we see in modern middle-aged human sexuality are artefacts, abnormal side effects of the situations and societies in which we now find ourselves. However, we know next to nothing about prehistoric sexual behaviour, so we do not really know how it differed from contemporary mid-life erotica.

However the data are interpreted, one thing is clear – nowadays, sex becomes less frequent in middle age. And this raises an important question: why?

Humans are unique in the animal kingdom in the multiplicity of reasons for which they have sex. While other primates, have sex for all sorts of non-procreative, social reasons, the sheer diversity of human sexual motives is staggering. As well as having sex to make babies, they also have sex to strengthen romantic bonds, to make them feel better or at least differently about themselves, to have fun, to distract themselves from pain or sadness, to rebel, to experiment, to make money, to express dominance or submission, to generate new routes of communication, to alleviate boredom . . . the list is almost endless. Because of this, humans often have sex at reproductively pointless times – at infertile phases of the female cycle, during pregnancy and even after the menopause. Perhaps eighty or ninety per cent of human sex takes place at times when conception is impossible – an expenditure of energy that would seem ridiculous in any other species (a female red deer may only have sex once a year, yet still produces one calf each year, despite her uninspiring sex-life). This bewildering complexity of the motives behind human sex makes it harder to investigate the middle-age decline in sexual frequency. Which of those motives are fading?

Especially complex is the relationship between bodily change and mental change. Many middle-aged couples who experience sexual problems do not know if these problems have a physical cause, a psychological cause, or both. And now we know why this is so – sex in humans is not a simple bodily activity driven by crude, predictable signals from the brain, as it is in many animals. Instead, human sex takes place almost entirely *within* that brain. Most of our reasons for having sex are emotional, social or psychological, so it should come as no surprise that it is so difficult to separate thinking from sex. Middle-aged people may say that their main sexual problems are physical (women: thirty-nine per cent vaginal dryness, thirty-four per cent lack of orgasm; men: thirty-seven per cent erectile dysfunction), but this is not necessarily true. Men are especially reluctant to believe that middle-age sexual change is anything other than physical, whereas women are more likely to acknowledge its mental dimension (for example, forty-three per cent of women cite 'reduced desire' as a cause of their sexual retrenchment). Men have a simpler, less introspective, more fatalistic attitude: they are more ready to accept the idea that middle age brings an inevitable physical sexual decline, and that this decline may make them feel less masculine. Women's concerns are more varied and include fears of becoming unattractive, worries about other aspects of life, and fears of ageing itself. However, they are also more able to accept that decreasing sexual frequency need not adversely affect their femininity.

Although I would argue that the physical side of sexual change is relatively unimportant in middle age, it does play a role. For a start, the clusters of cells deep in the brain which direct sexual behaviour in most mammals, the hypothalamic neurons, decline in size as we age. However, their importance in humans may be rather limited anyway, because the drive towards the 'mentalizing' of sex in our species has recruited other, larger, more complex and resilient brain regions into our mental sexual arena.

A second way in which sex changes in middle age is that sexual information entering the brain declines in quality. We receive sexual stimulation via all five senses, and earlier we saw how vision, hearing, smell and taste lose their sharpness in middle age. Also, touch is especially important for sexual arousal, and it experiences perhaps the most dramatic decline of all – our ability to discern fine tactile stimuli has halved by mid-middle age.

A third biological alteration which is often claimed to underlie sexual change in middle age is a decline in hormone levels. For example, it is likely that, on average, circulating levels of androgens (such as testosterone) decline gradually over a man's adult life, although certainly not as abruptly as the levels of sex hormones plummet during the female menopause. However, as we will see later, the decline is not a consistent phenomenon in all men, and may not be a bad thing, anyway – high androgen levels have also been linked to aggression, low achievement, and being rated by women as a poor sexual and marital partner in experimental surveys.

So the physical causes of the decreasing frequency of sex are overrated and far from universal. Many middle-aged people are very sexually active, sometimes more so than when they were younger, and they often express greater satisfaction with their sex lives. Most middle-aged people's sexual plumbing is in reasonable working order – certainly good enough not to be a hindrance – so it seems unlikely that the physical mechanisms of sex are inherently programmed to deteriorate beyond the point of usefulness during middle age. The 'clock of life' – the genetic programme which guides our development well into middle age – keeps these sexy mechanisms ticking over nicely even when couples have lost the ability to create children. Once again, this is clear evidence that human sex is for much more than just breeding.

Instead, it is the mental dimension of sex which is most vulnerable to change. Unlike the mechanics of copulation, the mental processes involved in sexual ageing are extremely complicated. When middle-aged people are surveyed about their reactions to sexual change, their responses provide a glimpse into the tortuous thought processes which preoccupy them. For example, middle-aged women often say they are worried that their partner will find sex disappointing because they have become unattractive, or because of a decline in vaginal muscular tone. Men rarely cite these as problems, but they do cite their partner's lack of apparent enjoyment and responsiveness during sex as a problem. Middle-aged men commonly worry about declining erectile function, yet the stress that this engenders can itself become the cause of just such a problem. And many women complain that it is not their partner's perceived erectile problems which worry them most, but rather the withdrawal from sex and intimacy which this perception causes. Clearly the brain copes less well with sexual ageing than the body does.

A dominant theme in the reduced frequency of sex in middle-age is our perception of ageing itself. Ageing is, of course, an inherently unfair process, and we all know that ageing tends to make people less attractive. Some of us still look good at the end of middle age, whereas some of us are distinctly frayed even by forty. And while studies suggest that facial attractiveness lasts longer than bodily attractiveness, both eventually take the same downward turn. Herein lie the three questions which many of us would rather avoid. Do middle-aged people have less sex because they find their partner less attractive? Do middle-aged people have less sex because they think their partner finds *them* less attractive? Or do they have less sex because they themselves feel less attractive, and therefore less in the mood for sex?

One of the reasons why these questions are unpalatable is that we all know that ageing is especially unfair to women. Everywhere we see examples of male actors and newsreaders having longer careers than their female colleagues, and this is borne out by studies which demonstrate that the advancing years have a greater effect on the perceived attractiveness of women than that of men (as assessed by both sexes). Wrinkles and greyness are more readily accepted in men than in women, but why? It is easy to claim that the western media have pushed us into this situation, but the cult of female youth is so dominant all over the world as well as throughout history that it seems more likely that it reflects preferences built deep into our brains.

Quite simply, evolutionary theory suggests that heterosexual women and men look for different things in a partner. This difference arises because of the degree of investment the two sexes put into their offspring. To raise a successful child (which is what is important in natural selection), a woman must invest a tremendous amount of time, energy and care – she has no choice. For a man to achieve the same end he has a range of options. He can invest a great deal in one woman's children, or he can spread his seed more widely and invest relatively little in each of his offspring, or he can steer a middle course and try a bit of both. Because females must invest so much, many animal species have developed a sexual ecology in which the behaviour of females is dominated by the availability of important resources such as food and shelter. And males do the same, except that for them the most important resource is females.

Because of these inherent differences, women across most human cultures seek out men who possess all the indications of genetic health (tall, handsome, masculine, symmetrical face and body) and signs that they are ready to invest in their children (status, intelligence, emotional commitment). Age is relatively unimportant because men remain fertile throughout their lives. Older men are, self-evidently, good at surviving, and they may also have accumulated more of a track record of resource acquisition, something which is by its nature attractive. There are limits to this attractiveness, of course, because men over sixty are likely to die before any child's requirements for resources have ceased. Also, very old men may have somehow reached their advanced age by investing less in important things such as breeding. Yet, however you look at it, middle-aged men are in the clear – theoretically sexy because they retain most of the advantages of youth, while also possessing the attractive features of age.

Men seek out their mates on a completely different basis. First of all, if they have no intention of investing more in a child than a few millilitres of semen, then theory predicts that they should be relatively uninterested in its mother's attributes. If, as is more likely, they intend to invest to some extent in their children then they would be expected to prefer women with signs of genetic health (pretty, feminine, symmetrical face and body) and potential to be reproductively successful in the future (curvaceous fat deposits, youthful breasts, youth in general) and sexually faithful. And this is where life gets unfair for women – men are not attracted to them because of their track record, but because of their potential: if a woman is young, she unarguably has more reproductive potential, more time left before the babies stop coming. And this is why, completely unfairly, people naturally view middle-aged women in a different light to middle-aged men. Indeed, some middle-aged women complain that they have become sexually and even socially 'invisible'.

As it happens, perceptions of masculinity and femininity also act to women's disadvantage during middle age. Masculinity and femininity may sound like vague terms, but I use them to mean the cues, often visual, which people find attractive in someone because they clearly define them as being a member of a particular sex. Masculine cues include muscle definition, a prominent chin, a broad chest, facial hair and a low-pitched voice, and none of these are much affected by the transition to middle age. Feminine cues include a round face with 'open' features, clear, smooth skin, thick hair, poor muscle definition, a waist, and subcutaneous fat on the limbs – in fact, they are many of the cues we also associate with youth. During middle age these feminine characteristics suffer direct attack by the ageing process.

However, although the prospects for the two sexes in middle age are starting to look distinctly different, there are other forces at play which may even things up somewhat. The first is that men probably use intelligence as an indicator of genetic fitness in women. As we have seen, cognitive abilities probably increase during middle age, perhaps especially in women, so this is one aspect of female attractiveness that one might expect to become more important. The second factor is that many human males choose the 'noble' option of investing heavily in the offspring of just one woman. Whether or not such 'monogamy' is the natural state for human males we will consider later, but any tendency for prolonged paternal investment in children would have led to a population of men more appreciative of the middle-aged women with whom they must spend much of their lives. The third force acting to 'even things up' between men and women is in some ways the converse of the second – that women may often behave in a more short-term way than men might expect. As we will see in a later chapter, there could be some logic in a woman who is approaching the end of her fertile years making a final, speculative attempt to have sex with a genetically superior male, even though she has no expectation that he will invest in her child.

So is it really loss of attractiveness which reduces the fre-quency of sex in middle age? I suspect not. It is noticeable that both men and women find middle-aged people of the opposite sex very attractive, and although there do not yet seem to be any studies to support this assertion I would guess that as people themselves approach middle age their tendency to find middle-aged people attractive increases. Twenty years ago I would not have wanted to be married to a forty-year-old, but today I am happy with precisely that situation. Speculating further, is it possible that middle-aged people undergo a specific mental change by which their romantic and sexual appreciation of their own partner is shifted, updated, so that they still find them attractive, despite the passing years? I do not know what the neural mechanism of such a change might be, but it would make a great deal of evolutionary sense. A couple can conceive a child in their forties and be caring for it until sixty, so it would be eminently sensible for their brains to self-rewire over that period so that the partner in whom and with whom they have invested so much remains, in their mind's eye, as alluring as ever.

It now seems clear that despite a partial (and inconsistent) decrease in the frequency of sex, middle-aged humans are surprisingly sexually active and attractive to each other, and to other age groups, considering how often they are dismissed as a spent sexual force. They are also free to enjoy one of the little gifts which evolution has bestowed on us. The sexual act is very brief in most animals – often less than a second in some mammals, yet in humans it can be extremely protracted and pleasurable – surprisingly so for an animal with so many natural predators just waiting to wolf it down in an unguarded moment. The leisureliness of human copulation says a great deal about its non-procreative, social, psychic purposes – human sex takes a long time because these other roles are so important. And of course the marginal waning in physical sexual function means that sex lasts longer in middle-aged people than in anyone else – and few of us would complain about that.

Yet the fact remains that sex does becomes a little less frequent in middle age, and an obvious effect of this is that we are less likely to conceive babies. But why should humans possess an inbuilt restraint which reduces their fertility in this way? This may seem strange, but as we will see, such self-restraint of fertility is not atypical – indeed, it is a dominant theme in the fifth and sixth decades of human life.

14. Why does women's reproduction just 'switch off'?

Menopause is fundamentally distinct from the reproductive senescence that has been described for other primate females.
Linda Marie Fedigan and Mary Pavelka,
The Physical Anthropology of Menopause, 1994

In our discussion of middle age, the menopause is the 'elephant in the room' which we have obstinately ignored up to now. At last, however, we can evade its tusks no longer.

The menopause has had an exceptionally chequered history in our culture, variously considered as a syndrome of 'oestrogen deficiency', a punishment for promiscuity or lack of education, a self-inflicted barrenness, a chastisement for failing to support one's husband and family, a disorder which requires medical intervention, or even in the famous words of one twentieth-century female scientist, a 'partial death'. Often there has also been a suggestion that men, too, should fear the menopause as callous nature's explicit reminder of their own deterioration and mortality.

Yet what interests me about the menopause is that it is biologically fascinating. In so many respects the menopause is bizarre, and in so many respects it confirms what I have been saying all along about middle age. It fits the distinctive-abrupt-unique triad perfectly – there is no other process so rooted in middle age, so temporally sudden, and so characteristic of our species. Although the experience of hot flushes and the implications of Hormone Replacement Therapy are important, to concentrate only on these aspects is to miss the biological point about the menopause. If we look at the natural history of the menopause, it tells us a great deal about the origins of our species. By challenging six erroneous assumptions we often make about the menopause, I hope you will start to view it differently.

Myth no. 1: Most female animals stop breeding as they get older

Should they ever think about it, many people probably assume that most ageing female animals stop breeding – that nature kindly protects elderly creatures from the exhausting strain of bearing and raising offspring. Yet as it happens, the pattern of human female reproduction, in which twenty-five years of generally good fertility are terminated by a rapid decline, is exceptional. For example, when scientists mollycoddle female farm animals so that they reach a wrinkly old age, they discover that although their ability to breed – regularity of cycles, chances of conceiving, vivacity of offspring – declines only gradually, there is no sudden end to their reproductive life. The sudden 'fertility cliff' seen in women simply does not happen.

One might expect the menopause to be more evident in animals which are closely related to us, and indeed, the evidence for the menopause in primates is somewhat stronger. In many female primates fertility does deteriorate with age and, just as in humans, a phase of erratic cycles and low fertility is eventually ended by a cessation of reproduction. However, this does not mean that these animals are undergoing a 'natural' menopause in the true sense – instead, we must view these findings in the context of how these animals live their lives. In non-human primates the end of female fertility is erratic, variable and often occurs in extreme old age. It is the end of a process of gradual deterioration very unlike the tightly timed, predictable process seen in women. And most important of all, fertility ends at or beyond the natural life expectancy of those species – in the wild, fertility almost never ceases. In contrast, the average age of the menopause in women is fifty to fifty-two years, and as we have seen, humans have often lived far beyond that age throughout our tenure on earth. So, unlike other primates, most female humans who reach adulthood will *naturally* undergo the menopause, and the menopause has added a new, long, post-reproductive phase to their life-plan which does not exist in our simian cousins. And of course, this unusualness requires an explanation.

If the menopause were unique to humans, then it would be difficult to study because we would have no point of comparison. Yet there are, in fact, a few animals which also undergo the menopause long before they die, and these include the killer whale and the pilot whale. In fact, post-menopausal whales may sometimes live even longer than post-menopausal women do. And we will return to these menopausal mariners later.

Myth no. 2: The menopause is when women stop having children

Women cannot naturally bear children after the menopause, but their ability to bear them has usually already vanished by the time the menopause occurs.

Irritatingly, the menopause can be defined only in retrospect – once twelve months have passed without menstruation, then the most recent period is defined as the date of the menopause. The reason for this contorted definition is that the last menstrual period is often preceded by years of erratic, variable-flow, skipped and prolonged cycles, when eggs often fail to be ovulated. All this time, the ovaries are shrinking as their valuable stockpile of eggs dwindles. Unlike the menopause itself, these pre-menopausal changes are gradual and halting. Women's cycles often grind to a stuttering halt.

Despite the fact that some of these changes start around the age of thirty, fertility is not much affected until forty. Thereafter, women's fertility declines to a point at which it is extremely difficult for them to become pregnant – mainly because the sheer unreliability of their cycles means they can no longer choreograph the interlinked biological processes of copulation, ovulation, fertilization and the establishment of pregnancy.

Later on we will explore recent changes in the age of motherhood in developed countries, but the story of women's fertility – their actual ability to conceive children – is the same across the globe. Fertility usually peaks around the age of twenty, but then declines only slowly until the age of forty. It is then that the real end of fertility occurs – in early middle age. The menopause may follow ten or more years later, long after childbearing has ceased (in eighty per cent of women the menopause occurs between the ages of forty-five and fifty-five). Despite recent trends, relatively few women have children after forty-three, yet the menopause often occurs after fifty. Thus, the menopause is not usually the time when fertility ceases – it simply confirms a cessation that has already happened.

Myth no. 3: The menopause happens when women run out of eggs

This is hard to explain. It is a confusing story and it is not helped by the fact that we do not actually know precisely how ovaries work. However, one thing is clear: there *are* eggs left in the ovary at the menopause. Not many, but they are there. And even to claim that this reduction in the number of eggs directly causes the menopause is misleading. What we wish we knew is why those few, apparently healthy eggs – potential babies all – stop working.

Let us consider the numbers. To ovulate one egg each month between the ages of fifteen and fifty, a woman needs approximately four hundred eggs. Of course no woman is that reliable – women get pregnant and stop cycling; cycles are erratic in the teens and after forty; some cycles lead to the release of multiple eggs or no egg at all; women suppress ovulation with contraceptives – but the four-hundred number is still worth remembering. Despite this low number of eggs which actually get ovulated, girls start life with a store of a few *million* eggs, for this is how many eggs a fetal human female carries in her ovaries. Yet by the menopause only a tiny fraction of that number is left (a few hundred?), so where did all the other millions of eggs go?

It is important to realize that very few of a woman's stock of eggs actually get ovulated – popped out of an ovary and into a Fallopian tube. Most eggs either start to develop and then fizzle out, or simply fizzle out without having developed at all – a process called 'atresia'. In fact, by puberty the number of eggs has already been whittled down to less than half a million, and all before a single egg has been ovulated. This process of atresia continues unabated throughout the next few decades and it is this, rather than the escape of a paltry few hundred ovulated eggs, which explains why there are so few eggs left by the menopause. In fact, if a forty-year-old woman could somehow arrest the process of atresia, she might still retain enough eggs to ovulate monthly for a further thousand years.

Thus, women do not really 'use up' their eggs, but instead there is a tightly controlled process of eggy death which carefully guides the number of eggs down to a level at which menstrual cycling fails. We are only now beginning to understand the genes and mechanisms which cause egg atresia, but it is clear that this process must be very precisely orchestrated. To take a pool of millions of eggs and trim it down so that it reliably reaches an unworkably low level at such a consistent age is a remarkable feat. To me, it does not look like the uncoordinated failure of an ageing body system – if it did, then women's fertility would stop much more erratically and in some women it might never stop at all. Instead atresia is a regulated, pre-programmed phenomenon – the careful paring away of eggs which precedes the menopause is part of the developmental 'clock of life'. It is this, and not the crude exhaustion of eggs, which determines the timing of the menopause.

Myth no. 4: The menopause happens *because* women run out of eggs

Even once the number of eggs is very low, this still does not directly cause the menopause.

Menstrual cycles are coordinated by undulating hormonal interactions between the nest of cells inside which each egg is sequestered (the follicle) and the pituitary gland which dangles from the underside of the brain. The pituitary secretes hormones called gonadotrophins ('gonad-growers') which stimulate ovarian follicles to grow and secrete their own hormones, including oestrogens and progesterone. These ovarian sex hormones then act reciprocally on the pituitary gland to moderate its secretion of gonadotrophins. However, something very strange happens at the menopause – the remaining follicles simply stop responding to gonadotrophins. They look healthy and the eggs inside them are presumably reasonably healthy, but the follicles simply do not secrete sex hormones any more, regardless of the amounts of gonadotrophins heaped upon them.

As a result, concentrations of ovarian sex hormones in the blood decline (progesterone by ninety-nine per cent, one potent oestrogen by eighty-five per cent, testosterone by twenty-nine per cent) and as the pituitary struggles valiantly to jump-start the ovaries, the concentrations of gonadotrophins in the blood increase spectacularly. These profound hormonal changes probably cause many of the 'symptoms' of the menopause which I will discuss later. However, the important point remains that we simply do not know why the ovarian follicles stop responding to gonadotrophins in the first place. What is it about having few remaining ovarian follicles which makes those follicles stop responding to gonadotrophins? This is a crucial question, because *this* is what causes the menopause.

In an attempt to find an answer, it is worth looking to see if there are any factors which control the timing of the menopause in healthy women. Smoking hastens the menopause, possibly by a direct toxic effect on follicles, whereas alcohol has no discernible effect. Some researchers have published evidence that good nutrition delayed the menopause in some nineteenth-century populations, and although these findings are not universally accepted, it is interesting to note that in modern developed societies high socio-economic status may also delay the menopause.

Then there are also several studies which indicate that ovulating less frequently delays the menopause. For example, women who spend more of their life pregnant or lactating reach the menopause slightly later. Also, women who had erratic cycles before the age of twenty-five, or who had long cycles, or who suppressed ovulation with contraceptives, all see similar small delays in the menopause. These results may seem unsurprising, but you should remember that it is not ovulation which depletes the ovaries of eggs, but atresia. So could it be that atresia, too, is slightly slowed when ovulation is suppressed? With this in mind, it is worth mentioning that many anthropologists believe that the pattern of reproduction observed in the developed world, in which women spend thirty years of their life regularly ovulating, is an entirely unnatural state of affairs. For example, Dogon women in Mali spend so much of their time pregnant or breastfeeding that they may ovulate only one hundred times.

The further one delves into the mechanisms of the menopause, the more convoluted they get – which is of course only fitting for such an important, sudden and distinctively human phenomenon. For example, the pattern of follicular loss may be more complex than just a gradual, guided descent into infertility. Counting follicles in healthy women is difficult, but there is now evidence that the slow follicular attrition of young womanhood is accelerated in middle age, as if a new mechanism kicks in to ensure that follicles are actively sacrificed to make the menopause occur 'on time'. Conversely, scientists have recently discovered that human eggs are not only lost, but also *made* during adult life: there are stem cells in the ovary which continually produce healthy new eggs and follicles. This replenishing of follicular stores may be a slow process, but theoretically women only need one egg per month to be fertile. And of course if these stem cells were shown still to produce eggs after the menopause, then this could have important implications for artificially restoring fertility in older women.

Another reason why post-menopausal women's fertility may be salvageable is that it takes more than ovaries to make a baby. For example, studies suggest that women's uteri retain the ability to support pregnancy even after the menopause, as long as they are supplemented with artificial hormones. On the other hand, the brain and pituitary seem to take an active role in hastening the onset of the menopause. In the years preceding the menopause it seems that the pituitary and brain appear slightly less responsive to hormones secreted by the ovaries, as well as 'pro-fertility' chemicals flooding in from elsewhere in the brain. As a result the pituitary may over-secrete gonadotrophins, which could damage the many healthy ovarian follicles still present at the start of middle age. Thus the menopause may be a cerebral phenomenon as much as an ovarian one.

Yet still the ovary confounds us. Why do the few follicles left at the menopause start to ignore gonadotrophins from the pituitary? How do they even *know* how many other follicles are left anyway? It is because of these questions that some researchers are now focusing on how the ovaries 'count' how many follicles are left – because this is what they seem to be doing. For example, there is a hormone made by follicles called inhibin, the levels of which decline as follicles are lost, and this decline has been implicated in the erratic waning of women's fertility in the decade preceding the menopause. Could ovaries use inhibin levels as some sort of aggregate 'count' of the follicles still remaining? Could this counting be what decides when the menopause happens?

One thing is clear: the menopause does not happen simply because all the eggs got used up.

Myth no. 5: The menopause is a universally negative experience

Although the effects of the menopause on the middle-aged female body are well known, what causes them is more mysterious. We assume that most changes are due to declining levels of ovarian sex hormones and increasing levels of pituitary gonadotrophins, but this assumption sometimes seems inadequate to explain what happens during the menopause, and why women vary so much in their experiences of it.

Some effects are relatively straightforward. Osteoporosis, for example, makes sense: oestrogens are known to maintain bone mass, and gonadotrophins may reduce it, so bones thin after the menopause and the likelihood of hip and vertebral fractures increases. In contrast, other changes often ascribed to the menopause may not actually be caused by it at all – and body shape is a good example of this. It may seem reasonable to assume that the changes in female body shape during middle age (for example, the 'loss' of the waist) are due to lack of oestrogens. After all, the development of the curvy, waisted female shape at puberty is driven by those very hormones. However, middle-age changes in body shape could just as easily be driven by the general mid-life trends of sarcopaenia and fat redistribution.

The menopause also affects the cardiovascular system. Hot flushes ('hot flashes' in the US, or my personal favourite euphemism, 'tropical moments') are a distinctive feature of the menopause, and are the menopausal change most apparent to other people. Indeed, many couples comment on how the man is the 'hot partner' during their twenties, whereas the woman takes over that role during middle age. We suspect that hot flushes may be caused by erratic patterns of gonadotrophin secretion, yet we really know very little about what causes them. Another unexplained effect of the menopause is that the tendency of women to suffer less cardiovascular disease than men declines, and this is often ascribed to hormonal changes. However, there is no sudden 'surge' in cardiovascular disease in women after the menopause, so the idea that oestrogens 'protect' the heart is perhaps simplistic. Maybe the narrowing of the gap between the sexes actually reflects a slight reduction in male cardiovascular disease at this age rather than any direct effect of the menopause.

Most controversial are the effects of the menopause on the brain. Despite what is often assumed, there is little evidence that depression is more common in women going through the menopause. In fact, the incidence of mental illness seems to be surprisingly unaffected by the dramatic hormonal changes taking place at this time. The menopause may have an impact on cognition, but this too is hotly debated. Surgical removal of the ovaries has been associated with a reduced ability to retrieve memories, yet despite the fact that some researchers have linked this to the actions of oestrogens on certain pathways in the brain, the effect of a natural, *spontaneous* menopause on cognition is unclear. As we have seen previously, the middle-aged brain is changing all the time, so any minor symptoms of the menopause could easily be swamped by this mêlée of cerebral change.

At first sight, the menopause appears to have a dramatic effect on female sexuality – before the menopause 60.7 per cent of women report having sex once a week; during the menopause that falls to 52.7 per cent; afterwards, it is 40.9 per cent. Yet if the data are analysed more carefully these trends start to appear more related to age than to the menopause – after all, post-menopausal women are older than pre-menopausal women. In fact, some studies suggest that the menopause has no direct impact on women's sexual desire or satisfaction. Many women actually report greater desire for sex, perhaps partly because they no longer have to worry about pregnancy and contraception (these findings come from studies of women who are not using any form of hormone replacement). Low sexual desire in middle-aged women seems to be more linked to stress, health problems and perception of their own bodies. And it is here that the menopause may have a minor effect, because if menopausal changes in hormone levels alter the middle-aged body, then they may affect middle-aged women's perceptions of themselves. As we have seen before, hormones often act on the human brain in indirect ways.

Considering how tightly controlled the biological process of the menopause is, it is surprising how much its effects vary. Some women suffer many severe adverse effects of the menopause, such as insomnia, altered mood, tiredness, severe hot flushes and possibly even memory loss, whereas others experience none at all. The adverse effects of the menopause vary between cultures, too – they seem to be worse in western developed societies, for example. Expectation is also important: the severity of symptoms correlates strongly with negative attitudes about the menopause expressed before it has actually started. This is even true of symptoms which appear entirely physical in nature, such as hot flushes. Given how consistent are the biological changes occurring within the body during the menopause, the variability of women's experience is mysterious but intriguing. Perhaps it results from the human tendency for so many aspects of our biology to be concentrated in the brain, which has gradually acquired spectacular dominion over the body during the course of our evolution. And we have already seen just how varied and individual our brains can be.

Although approximately one in ten women experiences significant adverse effects of the menopause, many women experience it positively – they may describe it as a beneficial change which freed them from menstruation, fertility and contraception, or as something of an anticlimax, worse in expectation than reality. Studies show that women's perceptions change as the menopause progresses – that they start to view it less as a medical complaint and more as a natural life-process. Also, women who already have children seem to have less fear of the advent of the menopause. In many human societies, the menopause is seen as a liberation from the risks of pregnancy and childbirth, and often leads to an increase in social status. In some cultures, the menopause is viewed as making women less of a threat to the stability of society. In others, it is seen as a reward for reproductive services rendered.

Myth no. 6: The menopause only happens because modern women live 'too long'

As we have seen, throughout our human history humans have frequently lived beyond middle age. And, as one of the defining features of female middle age, it is clear that the menopause is an essential and distinctive part of the human life-plan, built into us over millions of years of evolution. The nature of the menopause itself – its delicate control and consistently effected mechanisms – speaks of a process crafted by millennia of natural selection. It does not just occur because humans now live longer than they were ever 'meant' to.

So why did humans, uniquely among primates, evolve a naturally occurring menopause and the dramatic fertility decline which precedes it? We know that natural selection favours individuals who produce many healthy offspring, so it might seem strange that humans should evolve a system which *stops* them breeding. Why in the distant past did human females who stopped breeding end up producing larger broods of successful children?

One theory is that mammalian eggs have a fixed lifespan – that they have a 'sell-by date' of approximately fifty years. Beyond this age, it is claimed, the eggs have acquired so much damage that they can no longer be used, so it makes sense for women to stop ovulating them. I have doubts about this theory, however. First of all, hardly any mammals live beyond fifty, so the 'sell-by date' theory cannot actually be tested. Also, the theory works just as well backwards: eggs may not last beyond fifty years simply because there is no mammal which breeds beyond that age. Finally, the idea that breeding should cease simply because many eggs are damaged does not ring true. If only a few eggs were left intact, then it would still make sense for females to get them fertilized, just on the off chance that they might wring another healthy child from their deteriorating ovaries.

A second suggestion is that women stop reproducing to prevent them competing sexually with the next generation. This theory may sound unlikely – earlier we refuted a similar argument in favour of the evolution of death – but it is true that there is remarkably little intergenerational sexual 'overlap' in humans. Whereas most mammals pass from puberty into the major portion of their life (eighty per cent of it, perhaps) in which they can breed with individuals of any age, humans breed for only twenty years of their long, long lives, and rarely conceive babies with partners of their parents' or children's generations. Yet once again this theory is 'reversible': one could claim that lack of intergenerational breeding in humans is the result, and not the cause, of the menopause. Also, it is difficult to see why preventing inter-generational sex should benefit middle-aged people, unless of course it prevents them from luring potential sexual part-ners away from their own children's embrace.

The next theory, often called the 'Mother Hypothesis', is more convincing because it relates the evolution of the menopause to other unusual aspects of the human life-plan. Human hunter-gatherers survive (and pre-agricultural humans survived) very well in the wild because our large brain allows us to adapt to harsh and unpredictable environ-mental conditions. In other words, we are clever enough to eke out a living when other animals would die. Yet we pay two prices for having a brain that large: first, the birth process is risky for human females because of babies' large heads, and second, it takes lots of time and resources to raise our brainy children. Both of these problems are unique to humans. Thus, it is argued that human women evolved the menopause because at some point in life the risk of dying during child-birth and leaving their *pre-existing* resource-hungry children motherless became so great that it outweighed the advan-tages of having more babies. Better to survive with a few thriving children than to die leaving many to starve. This theory does make sense, and it also has some evidence to support it. Data from the pre-industrial USA as well as modern developing countries suggest that parents' longevity is reduced by each successive child they produce, and that this effect is much more dramatic for mothers (who undergo the menopause) than fathers (who do not). In this way, it seems there is considerable pressure for women to quit breeding and stick with the children they have. Maybe middle age is the time when this pressure becomes overwhelming.

There are two additional mini-theories which make the Mother Hypothesis even more convincing. First of all, if we assume that the likelihood of maternal death during childbirth, abnormality of offspring, and the incidence of stillbirth, all increase as women get older, then the advantages of continuing to breed get fewer and fewer, so employing the menopause to focus maternal care on previous healthy children becomes an even more attractive option. The second mini-theory relates to men. In humans, fathers often make a large contribution to the care and provisioning of their offspring, but their contribution is less reliable than that of mothers. Not only do men die younger and have more accidents than women, but it is not unknown for them to forsake their partners for other women. Thus, as a mother approaches middle age, the chances of the father of her children being present steadily decrease. This adds to her importance as prime carer for her children, and makes her continued survival even more crucial.

The most well-known explanation for the evolution of the menopause in humans is the 'Grandmother Hypothesis', presumably because it gives every grandmother who hears about it an opportunity to nod smugly. According to this theory, not only did natural selection favour women who stopped breeding and survived to rear their offspring to maturity, but it also favoured those who survived even longer – to help their daughters care for their grandchildren. Earlier, we saw that human offspring are so demanding that two parents are usually not a sufficient workforce to gather resources for them – other people must be recruited. Especially during lactation, women are unable to acquire large quantities of our hard-to-find food because they have lots of other tasks to occupy their time. In contrast, post-menopausal grandmothers are liberated from the inconvenience of breastfeeding and are free to forage. Also, there is good evidence that women's foraging efficiency increases well into middle age, just as it does in men. Some variants of the Grandmother Hypothesis emphasize the importance of grandmothers in directly caring for their grandchildren, but this is not necessary for the theory to work – gathering calories for those hungry grandchild-mouths is sufficient. It is almost as if grandmothers represent a non-reproductive caste of foragers who supply the needs of the reproductive few. In other words, grandmothers may play a role akin to that of worker bees.

Some demographic studies suggest that the presence of grandmothers does indeed increase their grandchildren's chances of survival. And, although computer models imply that the Grandmother Hypothesis alone is not sufficient to explain the evolution of the menopause, in combination with the Mother Hypothesis, it probably is. Other studies have looked more closely at the effect of grandmothers and have suggested that only *maternal* grandmothers have a beneficial effect – implying that only maternal grandmothers lavish the sort of care and resources on their grandchildren which actually benefits them in the long run, and that this discrepancy occurs because only maternal grandmothers can be certain that the grandchildren are theirs. In fact, there seems no limit to the Machiavellian deviousness of grandmothers, and it is now thought that they may bias their provisioning towards grandchildren who inherited more genes from them in the form of X chromosomes (because of the XX/XY system of sex determination in humans, the genes on the X chromosome are passed down unevenly from mothers and fathers to sons and daughters and grandsons and granddaughters). It seems increasingly likely that the effort which older women invest in their second-generation progeny depends on amazingly sophisticated subconscious assessments of genetic relatedness.

All of which brings us back to the whales. Pilot and killer whales really are very like humans in several ways – they are big-brained, intelligent, social, verbally communicative, and inventive in their food gathering strategies. And intriguingly, they are also among the few species which appear to undergo a human-like 'natural' menopause. Long-term studies of whale populations have shown that females spontaneously cease breeding long before they die – in fact, female whales' post-reproductive survival may be even longer than that of humans. This is important because it gives us an opportunity to compare our own biology with that of another, distantly related species. So do whales fit our theories of the menopause?

First of all, whale calves survive slightly better if their mothers are older – and this superficially seems to tie in with the Mother Hypothesis. Second, the presence of a grandmother whale appears to increase the chances of her grandchildren's survival, especially around the age of three. Yet although this sounds similar to the Grandmother Hypothesis in humans, it is unclear whether grandmother whales actually 'provision' their descendants. Also, although there is a tentative concordance between whale reproduction and our theories about the human menopause, it is important to realize that whales and humans, although similar in some respects, are not absolutely identical in their reproductive biology. For example, the mating arrangements of these species seem very different to those of humans – there are no long-term pair-bonds. However, whale social structures mean that females who live in one patch of ocean are often closely related to each other, thus making it advantageous for females to help nearby juveniles to which they are probably related (but how they help them, we do not know). Maybe they are even giving their grandchildren verbal advice – it has been claimed that old female whales may train young whales to avoid possible threats such as suffocating pack ice. Might there be a complex, chatty society under the sea, guided by a protective matriarchal culture?

So that was the menopause – or at least my Machiavellian-deviousness-gonad-grower-killer-whale version of it. What conclusions can we draw? That the menopause is 'almost unique' to humans, does not end women's fertility, does not happen when or because their eggs run out, is not as bad as we are often led to believe, and probably evolved so that middle-aged women could concentrate their efforts on their existing children and grandchildren without risking further perilous childbirths. And there is a wonderful upside to all this: women undergo the menopause because their offspring require an exceptional amount of care and support, and as a result human females live long, healthy post-reproductive lives.

Yet people argue ferociously about whether we should 'treat' the effects of the menopause. Proponents describe the menopause as an 'oestrogen deficiency' which requires a remedy, whereas opponents view the menopause as a natural process which should not be tampered with. Clearly some of the effects of the menopause can be alleviated by hormone supplementation – osteoporosis is an obvious example, although the doses of hormones required for this may be far lower than those administered in common HRT regimens. In contrast, the effects of hormone supplementation on post-menopausal sex life are controversial. And there are risks involved in HRT, which has been implicated in cancer, cardiovascular disease, gall bladder disease and even dementia, although these too are disputed.

Perhaps the most important thing women should consider is *why* they want to use HRT. Some studies suggest that the commonest reason is women's worries about their own attractiveness, and some argue that this is not sufficient reason to start taking artificial hormones. A purist might consider 'treating' the menopause to be an artificial distortion of an entirely natural process, but of course the same could be said of vaccinating against polio, fixing a broken leg, and not spending the rest of your life hunting and gathering on the African plains.

15. Crisis? What crisis?

There is one thing I want to make clear at the outset. There is one reason and one reason only why I bought a Lotus. I did not buy it to alleviate angst about my physical or mental deterioration, or to plug leaks appearing in my ageing psyche, or to salve my awareness of impending mortality, or to ensnare young women miraculously attracted to ageing chubby men in bright blue cars. Instead I bought a Lotus because I have wanted one ever since I was eight years old. The fact that I actually *did it* at the age of forty-one is mere coincidence.

People love the male mid-life crisis. It always raises a smile. Yet I hope to convince you that it does not actually exist. In fact, it never really existed in any concrete form. Initially popularised in the 1970s in the US, it supposedly took place at a vaguely defined time in life, and seemed to involve a three-way mix of worry about bodily deterioration, pathetic desire to seek the romantic attentions of younger women, and urges to indulge in childish activities, all seasoned with an overarching sense of 'taking stock' and psychological malaise.

Yet the male mid-life crisis is not real. If it were, I could have justified saving up for an Aston Martin.

*

Of course the male human body does change during the fifth and sixth decades of life. I do not wish to hark back to the grey days of Chapter 5, but there is evidence that some parts of men's bodies are subject to slight deterioration.

Balance fails, blood production slows, immunity wanes, sleep becomes erratic. Yet bone, muscle and sinew are what men consider important. We have already discussed how muscle mass decreases in middle age – 'sarcopaenia' – and how this exerts more noticeable effects on appearance and strength in men than women. Bone density peaks in the late thirties and gradually declines thereafter, although the diameter of male bones increases to compensate for this. Loss of bone density is a consistent finding in all human populations, but does not become dramatic until after the age of fifty. The risk of hip fractures increases, but they are approximately three times less frequent than they are in women simply because men's bones are more robust to start with. Yet it is still important to realize just how important these musculoskeletal changes are – middle-aged men's most commonly reported health problems are joint pain and back pain, rather than anything more intimate or psychological.

As well as being painful, these musculoskeletal changes also alter men's appearance – posture alters, foot arches sag, spines bow and the discs between vertebrae are compressed – and as a result middle-aged men get shorter. Also, because the loss of height is due to changes in the spine, rather than shortening of the limb bones, this accentuates the middle-aged fat-with-spindly-legs appearance I mentioned earlier. On average, men get shorter by one millimetre per year after the age of forty, although some may eventually lose as much as seventy millimetres. Height is important to men, and this may partly explain why their body image deteriorates in middle age – in fact, it deteriorates faster than women's, so the gap between the sexes is narrowed. However, middle-aged men still retain elements of their in-built confidence – for example, although they are more likely to diet than when they were younger, they are still less likely to diet than women. Also, just as in their youth, middle-aged men still overestimate the male body weight which women find desirable (women consistently *under*estimate men's ideal female body weight). And the final refuge of the male middle-aged paunch is that its bearer can claim it is visible evidence of success in life, as well as being a useful target for humour. Middle-aged male bellies are inherently funny, and thus perhaps subsume the comedy role played by the penis in earlier years.

In contrast, sex is a serious business for middle-aged men, although here the changes are subtle. The data are not superficially encouraging, I admit – frequency of sexual day-dreaming decreases, nocturnal erections become less frequent or prolonged, erections are achieved more slowly, manual as well as visual stimulation becomes more necessary, orgasms are shorter, ejaculatory volumes smaller, erections subside more quickly, and the 'refractory time' until another erection can be achieved increases. However, a great deal of human sex 'happens in the mind', so the effects of these physical changes can be surprisingly small, and they are often obscured by positive or negative changes in men's 'cerebral' sex lives. Also, there is no discernible change in the fertilizing ability of men's sperm during middle age, and perhaps most importantly, the increased time required to reach orgasm is often welcomed by all concerned.

There is controversy about what happens to men's sex hormones during middle age, whether they are responsible for changing their sex lives, and whether they require supplementation. Some have undertaken an aggressive search for a male transition as clear-cut and consistent as the female menopause, and have variously labelled it the 'male menopause', 'andropause', 'partial androgen deficiency', or the menacing 'late-onset hypogonadism'. Yet gradually the tide has turned against the idea that men's reproduction declines as suddenly as women's.

Concentrations of androgens (testosterone and similar hormones) in the blood do, on average, decline with age. This may start long before middle age, perhaps as early as twenty, but in middle age the average decline is perhaps 1.0 to 1.6 per cent per year. The decline occurs partly because the androgen-secreting Leydig cells in the testicle decrease in number, and this androgen decline leads to increased secretion of gonadotrophin hormones from the pituitary gland (because the testicles and pituitary control each other in a similar way to women's ovaries and pituitaries – see the last chapter). There is evidence that the average decrease in androgen levels may accelerate after the age of fifty, which would place it intriguingly close to the average age of the female menopause, but probably too late for the supposed mid-life crisis.

However, this tendency for average androgen levels to decline does not look much like a consistent, major turning point in middle-aged male life. Many middle-aged men's androgen levels remain unchanged well into old age, yet in some of these men libido and potency still decrease. And conversely, other men experience an androgen decline while experiencing few apparent adverse effects. In general, the links between androgen levels, sex life, disease and general bodily deterioration in men are extremely unclear. Sperm production rarely shuts down, and the testicles almost never stop responding to pituitary gonadotrophins (women's ovaries *always* do). Indeed, one large US study suggested that only one in fifty middle-aged or elderly men shows evidence of complete androgen 'collapse'.

Thus, 'the andropause' is slow, not restricted to middle age, erratic, strangely variable between individuals, and has no definite effects. Also, it does not involve complete failure of the interplay between brain and gonads. In other words, it is completely unlike the female menopause. Yet still the arguments rage about whether we should 'treat' it or not. Hormone supplementation has not been shown to improve any aspect of middle-aged life in the vast majority of men, and may in fact increase the chances of diseases such as prostate cancer. And once again, the human brain seems relatively unaffected by hormones: androgen therapy has surprisingly little effect on libido or sexual behaviour.

One thing which many middle-aged men fear is erectile dysfunction, yet even this complaint seems little affected by hormones. The mechanisms of human erection are becoming increasingly well understood, mainly because of the enormous potential profits available to anyone who invents a drug which encourages male tumescence. During erection, a complex mix of chemical transmitters rains down on regions of the brain which send nerve fibres down to the genitals. These nerves stimulate the release of nitric oxide from the inner lining of blood vessels in the penis, and this causes those blood vessels to relax, allowing the spongy erectile chambers to engorge with blood. This, for example, explains how drugs such as Viagra work – they inhibit the activity of an enzyme ('cyclic-GMP-specific phosphodiesterase type 5') which suppresses the chain of events which leads to erection.

Also, the central role of blood vessels in erection explains why erectile dysfunction often occurs alongside atherosclerosis, obesity and diabetes, and why it may be exacerbated by lack of exercise and smoking.

Just because we understand the physical basis of erectile dysfunction, we must not ignore its psychological dimension. Men worry about impotence, and erections are inhibited by worry, so erectile dysfunction is often largely psychological in origin. It is striking that many men with erectile dysfunction undergo normal nocturnal and morning erections – clear evidence that the flesh is willing but the spirit is weak. For example, something as innocuous as mild boredom can cause men to embark on damaging cycles of erectile failure, self-doubt and anxiety. Even impotence which has a clear physical cause is often exacerbated by its effects on men's sexual self-confidence. Men often assume that 'failed' sex is their fault – and of course it is an inescapable biological fact that copulation is indeed more dependent on the arousal of males than females. 'Failed' sex has dramatic effects on men's views of their own sexual identity – more so than women, according to psychological studies. Worse, men also worry that their partner will interpret 'failed' sex as a tacit comment on declining female attractiveness, and indeed studies show that this is precisely what many women *do* think. Cycles of anxiety in men; cycles of anxiety between partners – the brain is humans' most important sexual organ, and nowhere is this more obvious than in erectile dysfunction.

So the gradual sexual change of male middle age is entirely unlike female middle-age sexual change. Spectacular declines and sudden cessations rarely take place. Instead, male reproductive function changes slowly – no faster than any other body system, in fact – and the brain's reaction to these changes is probably more important than the capricious ebb and flow of hormones. Men's fertility declines gradually and partially with the accumulating years, and they are spared a sudden onset of infertility. The abrupt male menopause is a myth, and soon we will see why evolution has made it so.

Perhaps the biggest cliché about male middle age is that it induces a sudden desire to impress younger women, and possibly leave a middle-aged spouse to run off with one. This idea exerts such a hold over us because it engenders anxiety in middle-aged men's wives, and even guilt in middle-aged men who find the occasional younger woman attractive. Yet is it a real, distinct phenomenon which suddenly strikes in middle age?

Of course, a simple challenge to the younger-woman theory is arithmetic – a man simply cannot lust after adult women twenty years his junior until he hits forty, because until then there *are* no such women. However, there must be *something* going on. Modern society does not frown so much on a seventy-year-old man asking a fifty-year-old woman out on a date, so why do we pick on middle-aged men so much? Let us look at the evidence.

Two things are consistent across all contemporary human cultures. First, women's attractiveness, as assessed by others, peaks in the late teens or early twenties, and then declines with age. Second, there are consistent age gaps between heterosexual romantic and sexual partners throughout the world, and similar trends are reported in historical texts. Although the age gap may have narrowed recently in developed countries, older men still seem to pair up with younger women. Also, in many societies, men habitually overstate their own age and understate that of their partner, whereas women often understate their own age, and overstate that of their partners. So consistent and ubiquitous are these findings that anthropologists claim they are not some trivial result of social mores, but instead reflect an inbuilt human sexual strategy – part of our developmental-genetic 'clock of life'.

At the time when men marry, they usually say they prefer women who are between two and seven years younger than them. In the UK, the average marital age-gap has been stuck between two and three years for the last century – and that average has not budged even though the *variation* in the age gap has increased (more marriages between much older men and much younger women, and more marriages between older women and younger men). Also, intriguingly, one study showed that the number of offspring produced by a couple was maximal if a husband was six years older than his wife.

Unfortunately, marriage data only tell us about the partners people can *get*, but we also need to know about what partners people *want*. One approach has been to study the preferences expressed by people placing ads in newspaper lonely hearts columns, or engaging in online dating, and the results are fascinating. At the age of eighteen, men seek women who are older than them, but this desired age gap gradually declines until, at twenty-four, they seek women who are similar in age to themselves. From then on, they seek women who are increasingly younger than themselves, until at seventy years old they seek women who are on average sixteen years their junior. This is a neat and powerful result – pretty much a straight line on a graph of desired partner-age. However, women do not seem to reciprocate, preferring in their late teens to be four years younger than their partners, but steadily reducing this desired age-gap to zero by the time they reach old age. (Incidentally, gay men show remarkably similar trends to straight men, except that their preferences are more extreme – wanting 'more old' partners when they are young and 'more young' partners when they are old.)

Of course, we must be careful when interpreting these findings. First of all, one could argue that they too demonstrate not what partners people want, but what partners people pragmatically think they might be able to attract. Would seventy-year-old men ask for twenty-year-old women if they thought their wish might actually come true? Also, the sample is selective, consisting mainly of people who are not in a relationship and who are (hopefully) not married. Yet the evidence still points in one direction – most men like younger women, and most women like older men. And indeed, the data chime with what evolutionary biologists tell us: that in a pair-bonding species with offspring who take a long time to grow up, males should be attracted to females with many years of potential mothering ahead of them. Men who did this during our evolutionary history – who were attracted by the cues of female youthfulness, who hitched up with young women, and who made the most of those women's fertile years – were the ones who ended up passing on their genes to the current generation of men, who are thus genetically programmed to do the same.

So, according to this theory, running off with younger women is not 'narcissistic refuelling' for middle-aged men, but an eminently sensible strategy. Indeed, there is a counter-example which adds weight to this argument. Chimpanzees' offspring take less time to grow up than ours, and chimps are promiscuous and do not pair-bond, so their reproductive life is very unlike ours. And male chimps actually seem to prefer to mate with *older* females. Whatever the reason for this (perhaps older females have demonstrated a track-record of raising chimp babies?), the contrast with humans is stark, and strongly suggests that our evolutionary theories about human male mate preferences are correct.

Yet does all this entirely explain male middle-aged behaviour? First, although young women may often strike middle-aged men as attractive, the urge to seduce them does not seem as overpowering as the theory might lead us to expect. Whatever some people might think, the world is noticeably *not* full of priapic middle-aged men chasing twenty-something sirens. Also, it is hard to explain why, before the age of twenty-four, men prefer older women, and only thereafter do the evolutionarily 'sensible' thing, and switch to desiring younger women. After all, in the ancient, pre-agricultural world it is likely that women started to produce babies around the age of sixteen, so why are men not programmed to make the most of sixteen-year-old girls' reproductive potential? Surely middle-aged men should all be craving sixteen-year-olds rather than women in their twenties who are, in comparison, rather over-the-hill?

Still, there is always potential for men's genetic inheritance and continuing fertility to tempt them from their middle-aged romantic union into trying their luck with a younger, more fertile partner. Yet this is far from a universal occurrence, and middle-aged men usually stick with their long-term partners to complete their parental labours alongside the person with whom they have already invested so much. And if that drive is not sufficient, the idea of re-entering a nightmare world of dating, nappies and teething is often enough to steer them back to the straight and narrow.

A completely different theory of why middle-aged men

supposedly want to run off with younger women relates to how they perceive their own mortality. It could be crudely phrased as 'get rid of the ageing wife and get rid of ageing'. It assumes that men use their partner's vitality as an indicator of their own, and indeed men do spend more time looking at their partners than looking at themselves. This may sound a strange idea, but there is evidence to support it. First of all, remarriage statistics suggest that middle-aged men tend to marry a second wife who is younger than their first wife (although remarrying women show similar trends). Also, men with younger wives live longer than men with older wives, and increasing age disparity enhances this 'protective' effect of a younger wife. We do not really know why this happens – whether a young wife has psychological, social or physical effects on her husband – but husbands' ages do not have such dramatic effects on their wives' survival. Yet before we take all these data as support for the politically incorrect adage that 'you are only as old as the woman you feel', there is one more piece of evidence to consider. This is that men's longevity also correlates strongly with their wives' level of education – so marry a smart woman and live long. But which is more important – spousal age or spousal education? In other words, at what point does it become more advantageous to divorce your intelligent middle-aged wife and run off with a twenty-year-old airhead?

And of course, it takes two to tango – in life as in evolution – so what about the younger woman who tempts a middle-aged man away? What is in it for her? Is it wealth, security, a man's proven track record of caring (although apparently not caring enough)? Some evidence suggests that it is in fact women's choices which are the dominant driver of human romantic couplings, and that all men can do is comply. In dating ads, women tend to proffer information about their youth, beauty and playfulness, whereas men give information about their status, wealth and employment. Yet some studies suggest that women are actually more choosy about prospective partners' age than men are, so how much choice does all this leave for middle-aged men?

Wherever the balance of power may lie between middle-aged men and younger women, there is almost certainly one beneficial result of these often vilified relationships. Over our species' history it is very likely that a reasonable proportion of middle-aged men sired children with younger women – either furtively, or after separation or bereavement. And as a result, all human males have inherited genes which keep them fertile beyond forty, on the off chance that they might just get lucky. Once again, natural selection's feminist credentials have proved unimpressive. In human male sexuality, it is not over until the fat lady has the last baby, and this is why men do not undergo a sudden male menopause. Indeed, these occasional ancient sexual shenanigans by middle-aged men have probably been a major force ensuring human vitality long beyond the age of forty. Thus, we can partly thank lecherous prehistoric middle-aged men for our long, healthy lives.

The third element of the mid-life crisis myth – along with the male menopause and younger-woman-seeking – is that middle-aged men undergo massive psychological upheaval which results in a novel and hilariously inappropriate approach to life.

According to this idea, middle age brings with it an internal re-evaluation of the self, a realization of impending mortality, a panic about lack of personal achievement, an identity crisis and an unwillingness to 'just do more of the same'. This supposedly leads to depression, anxiety and non-sensical responses – ignoring problems, regressing to an adolescent mindset, substance abuse, divorce and even suicide. The suggestion is that middle-aged men make one last desperate effort to turn themselves into heroes, and usually fail, sometimes with disastrous consequences. Most important of all, there is remarkably little evidence for any of this.

For the mid-life crisis theory to work, the crisis must be quite consistent between individual men, and it must also occur over a relatively short time-span. However, only one in ten men is reported to experience intense emotional turmoil in early middle age, and as we have seen diagnoses of mental illness, including depression, do not become more frequent in middle age. When asked about major turning points in their lives, most men mention events in early adulthood rather than events in middle age – career changes and mar-riage, and perhaps education. To a great extent they see their young adulthood as the phase which set the tone for the rest of their lives. Of the men who do believe that they underwent a mid-life psychological crisis, more than half say that it occurred before the age of forty or after the age of fifty. And of 'crises' which actually do take place in the forties, most have an obvious external cause, such as the loss of a job or a marriage breakdown, rather than any clear link to any partic-ular chronological age. In fact, in some surveys women were just as likely as men to say that they had undergone a mid-life crisis.

So not only is the definition of the mid-life crisis pitifully vague, but the evidence for it is even vaguer. Men do report psychological changes in mid-life, but the changes they mention suggest that they are more concerned about failing cognitive abilities than any major intrapsychic emotional reorganization. The commonest complaints are loss of concentration, tiredness, irritability and reduced memory – a list which could reflect perceived failings rather than real ones, and in any case hardly betrays a huge groundswell of middle-aged male angst. In fact, the only piece of evidence which shores up the incoherent fiction of the mid-life crisis is that there is a very slight, transient upturn in the frequency of perceived 'life turning points' during middle age in men (and not women). Once again, though, many of these are career-related and many of them are viewed as positive changes, rather than negative ones.

The current consensus among psychologists is that the male mid-life crisis is an ill-defined, evidence-less concept which has outlived its usefulness. But if this is so, why do we still cling to it with such enthusiasm? I suspect that we like the mid-life crisis because it tells us something we want to hear. Sometimes, it may even have romantic, heroic overtones. For example, great thinkers and doers of history (who were, for various cultural reasons, usually male) have often been reported to have undergone a dramatic mid-life shift in their perception of the world. Whether or not Michaelangelo Buonarotti, Johann Wolfgang von Goethe and Dante Alighieri would agree with posterity's appraisal of their middle age, the idea that a sudden intimation of mortality can drive a man to great things is an attractive one, especially when most of us are not sure if *anything* could drive us to great things.

No – most of us, rather than reworking human culture, just work. And the world of work changes all the time for middle-aged men. Although the pressure on middle-aged men to provide for their families has now been partially relieved by the entrance of many of their partners into the workplace, the world of male work remains a complex one. The post-war lifelong pact between male worker, employer and welfare state is breaking down. The post-baby-boomers are a swollen generation in a shrunken economic world, and financial planning for old age is becoming increasingly unpredictable. Some men can retire early but some must retire late, all the time angered by the fact that the early-retiring individuals are often the ones who spent decades trying to down-size them out of the workforce. In short, there is no simple biological theory to explain the predicament of middle-aged men in modern economies. Crises may come from without, but they certainly do not come from within.

And what of the adolescent behaviour of middle-aged men? Why, exactly, *did* I buy a Lotus? Psychology does not tell us a great deal about this issue, except that there is no scientific evidence of infantilization in the middle-aged male psyche. Could it be that middle-aged men are not reverting to their childhood, but rather that their adulthood was never as different from their childhood as they might have wanted us to think? Apart from acquiring a novel interest in girls around the age of fourteen, men's interests do not appear to change very much after the age of three. They like physical exertion and competing and wheels and machines and making things and destroying things and generally acquiring and fiddling with *stuff*. Puberty and romance are but a brief interruption to this, and having one's own children only confirms it. Now I have a son (sons are a great excuse for shame-free middle-aged male play) I have rediscovered, or perhaps come clean about, my love of Lego. I like playing with Lego now in the same way I did when I was five – it has assumed no new mature subtleties for me – and I suspect that many men would say the same thing. I believe I would have liked playing with Lego when I was twenty as well, but no doubt I temporarily set it aside to avoid social embarrassment and romantic failure. And now, like many men, I have a little more cash, a little less pride and a suspicion that now is the time to be a little more selfish and realize some of the plans that five-year-old-me once made. Once men have grown older, there is at last time to be young. Hence the Lotus.

One final aspect of the crisis myth is so pervasive that until someone points it out, you may not even notice it. In our politically correct, post-feminist, non-ageist world, there is one group of people who may be mercilessly lampooned without fear of reprisal, and that is middle-aged men. Next time you watch television you should make a conscious effort to see who is being belittled in the comedies, dramas and commercials. Most of the time it is middle-aged men playing buffoons, benevolently tolerated by their wise, intelligent, articulate wives and children. Middle-aged male characters are often physically unattractive, too, to an extent which makes the casting of their usually beautiful screen-wives strain credibility. I am not referring to the wonderful comedic power of surreally, *intelligently* stupid middle-aged men like Homer Simpson or Peter Griffin, but the insidious drip-drip of lazily penned minor middle-aged male characters who ask no more than to repeatedly require condescending rescue by their long-suffering middle-class families. Of course, middle-aged men are media fair game because it is they who for so long wielded the power in society, but I cannot help feeling that their representation in the contemporary media only feeds the myth that something is fundamentally awry in the middle-aged male.

*

Most of all, the mid-life crisis is a great story. We all like stories, and it does not really matter if they are fiction. Perhaps it makes them even better. The mid-life crisis is such a neat, bite-size concept that it is tempting to believe it must be real, even though it has no psychological basis. It is convenient and of course it is funny – so much so that even men themselves like talking about it. And to make it even more attractive to men, it is an abstraction – a concept into which any minor upheavals and fears may be neatly packed and locked away, hidden from view in a humorous receptacle. Some have suggested that the mid-life crisis developed in recent generations as an apologetic substitute narrative for men who never suffered in a war or a depression – implying that today's men are somehow embarrassed by the lack of real, manly suffering in their lives, and cover this embarrassment with a fictitious story called the 'mid-life crisis'. But do we really need such theories to explain why people find it comforting to laugh at middle-aged men?

The author doth protest too much, I hear you say. Perhaps an early-forties male is not the best person to be making this case. Maybe denying the mid-life crisis is just my personal way of coping with it. Anyway, I would now like to forget the mid-life crisis and move on. After all, as all men know, there is *stuff* to do.

16. Should middle-aged people have babies?

Right from the start I was special. Not only was I born bottom-first into the glare of twenty-or-so medical students, all gawking at the only such freakish occurrence during their obstetrics rotation, but I was also born to a mother who was officially old.

Nowadays, thirty-one may not seem very old to have your second baby, but in 1968 it was apparently viewed as being towards the senile end of motherhood. As a result, my mother was isolated at one end of the maternity ward, in an area reserved for 'older gravidae'. Perhaps the doctors were worried that she might infect the other mothers with *age*. The idea of thirty-one being an advanced age to produce a child seems quaint today, and parenthood has changed a great deal in the time it has taken that product of 1968 to reach his own middle age. We all know that many people now start families much later, and in this chapter I will tackle a crucial question about this trend. In the third part of this book we have already examined sex, female fertility and male fertility in middle age, but assuming that these forces conspire to produce a baby, is middle-aged parenthood actually a good thing?

The numbers are clear. In the UK, between 1989 and 2009 the number of women over the age of forty having children tripled, while the number of women under twenty-five having children decreased. In the US, during the decade between 1991 and 2001, the number of women over forty-five having children went up by 190 per cent. There is no doubt that this is a real phenomenon, but no two families' stories are alike. Many middle-aged women conceive naturally, but some require the help of modern reproductive technology to conceive their own children, while others must resort to egg donation. And of course, more older mothers also means more older fathers – something which is too often forgotten, especially as some older men also spend their middle age starting second families with new, younger wives.

Middle-aged mums are not a new phenomenon, but they are not the same as they used to be. Until the last few decades, most middle-aged mothers were women who had procreated ever since marriage, so their middle-aged pregnancies were simply the completion of large, protracted and presumably exhausting families. What has changed recently is that more parents are *starting* their families in their late thirties or early forties. Also, there is evidence that these mothers are in a rush to beat the biological clock – the interval between first and second births is shorter in middle-aged mothers than in younger mothers, even though they are less fertile. This is clear evidence that most middle-aged pregnancies are conceived (if you will excuse the pun) as part of a plan: a plan to sneak in a few offspring before the uncaring guillotine of infertility falls. Modern middle-aged pregnancy is not a biological innovation – it is a social trend.

Demographers have told us what has happened to childbearing in developed nations. Now women's socio-economic status is more defined by what they themselves do, rather than just by whom they marry, they face conflicting priorities. In short, women face a choice between either reproducing early and sacrificing socio-economic success, or establishing a successful career as a high-earning, flexible worker, and then taking time out to produce a family when they are in a strong position to negotiate a career break. Money seems to play an important part in these decisions – before the birth of a first child, a husband's income has little effect on a woman's activity in the job market; but once that child is born, then the more that husband earns, the less likely the woman is to re-enter the market. Financial and career stability is often seen as a prerequisite for starting a family, and modern economics make it hard to achieve this until well into the thirties. Also, increasingly lengthy education and training is now required for women even to start the careers which offer them their desired future salaries, and even once those salaries are attained, the prospect of temporarily forgoing them is a powerful disincentive to taking a career break. All in all, it is hardly surprising that middle-aged first-time mothers tend to be well educated and financially successful.

There has been a great deal of discussion about these trends. Some even worry that they make successful women less likely to contribute genes to the next generation. Whatever the truth, it is clear that money and possessions have changed everything. Before the advent of agriculture, our only wealth was the fat carried within our bodies, so our brains evolved to make the relatively simple decision about when to convert that adipose wealth into babies (probably at every available opportunity, as it happens). Yet most of our wealth is now non-biological, outside us, and slow to accumulate. Is it any wonder that modern economics and female choice have led us into a dilemma to which our brains have no innate, instinctive solution?

Despite the recent trend, there remains a residual unease about older mothers, but few of us can explain it, or even be precise about it. If we think that pregnancies in thirty- or forty-year-olds are fine but find ourselves baulking at pregnant sixty-year-olds, where do we draw the line? And why *do* we question the notion of pregnancy in older mothers? Is it because there is something truly wrong with it?

From a biological point of view, there are four potential disadvantages to middle-aged parenthood, and the first of these is that it might be a risk to the mother. We have already seen that women's fertility usually declines dramatically in early middle age – the forties – but does this mean that pregnancy, if initiated, is necessarily risky?

In fact, the evidence is equivocal. Some studies clearly indicate that older mothers are more likely to suffer pregnancy complications such as gestational diabetes, placenta praevia, pre-eclampsia and hypertension, or to undergo Caesarean section. However, in some of these reports the study group consisted of women over fifty – an age at which most women are unlikely to conceive anyway. Even if such risks extend down into the forties or beyond (some claim that they extend down as far as thirty-five), then they may still merely reflect the effects of diseases already present before conception. In other words, even if age per se does not increase the risk of pregnancy complications, the fact remains that middle-aged women have had more time to accumulate the pre-existing medical conditions which can precipitate or exacerbate such pregnancy complications. Thus it remains unclear whether a healthy middle-aged mother has anything to worry about.

The second potential disadvantage of middle-aged parent-hood is that maternal age may adversely affect a baby's health, and of course the condition which dominates public debate of this issue is Down's syndrome.

Down's syndrome occurs when a child inherits an abnormal complement of chromosomes. Most children inherit forty-six chromosomes – twenty-three pairs, with one of every pair coming from each parent. The chromosome pairs are numbered from chromosome 1 (the largest) to chromosome 22 (the smallest), with the final, twenty-third, pair being the sex chromosomes (two Xs in girls, and an X and a Y in boys). For reasons we do not understand, the smaller chromosomes are less likely to undergo the normal shuffling, rearrangement and splitting which is essential to make an egg cell in the ovary, with its demi-complement of twenty-three chromosomes. As a result, some eggs end up with not one but two copies of a chromosome, and if these eggs are later fertilized and the father supplies his usual single copy, then the resulting child will have three copies of that chromosome, and thus forty-seven chromosomes in total. This is called 'trisomy', and the commonest trisomies are those involving chromosomes 21 and 22; trisomy 21 causes Down's syndrome.

We do not know why it is the smaller chromosomes which are more at risk of trisomy, nor why trisomies of larger chromosomes seem different (16 is the next most common, but follows different rules). Yet two things are clear – trisomies almost always result from an error in the formation of the egg, not the sperm; and trisomies are more common in babies born to older mothers.

This is an important issue – chromosomal abnormalities

('aneuploidies') occur in approximately 0.3 per cent of newborn babies, four per cent of stillborn babies and up to thirty-five per cent of spontaneous miscarriages. Yet it is important not to leap to conclusions about the role of older mothers in chromosomal abnormalities – especially as those conclusions can lead to a damaging mix of fear and guilt in middle-aged women. For example, it is often assumed that egg aneuploidies occur because middle-aged women's eggs have been accumulating damage for over four decades, and thus are inherently inferior. However, recent studies have shown that the ovaries of female human fetuses already contain large numbers of aneuploid eggs, so chromosomal abnormalities are not a direct result of ageing. Instead, it is possible that these abnormal eggs are for some reason more likely to evade the process of egg attrition which we have already seen scythes the fetal store of millions of eggs down to the few hundred that will eventually be ovulated. Perhaps so much scything has taken place by middle age that the preferential survival of aneuploid eggs means that a forty-year-old's pool of eggs has effectively been 'enriched' for abnormal eggs.

The figures can certainly look alarming for prospective middle-aged mothers – the incidence of Down's syndrome increases at an ever-accelerating rate throughout a woman's adult life. And the incidence of Down's syndrome is so very low in twenty-somethings that by the time its incidence reaches 1-in-100 at the age of forty it can seem, in comparison, frighteningly high. Yet it is important to step back and examine these probabilities carefully, and from an individual's point of view. Is one per cent such a terribly high risk for the most common human aneuploidy? Of course, it is up to every prospective parent to decide whether this level of risk is acceptable – and to bear in mind that Down's syndrome is usually diagnosed early in pregnancy, so parents can now decide whether or not to bring a child with trisomy 21 into the world.

Although it is difficult to obtain precise figures, it is clear that miscarriage becomes more common in older mothers at a rate which also accelerates with the passing years. Women probably pass the point at which they lose half of their pregnancies some time between the ages of forty and forty-five, although the numbers are uncertain because losses in early pregnancy may go unnoticed. Thus, miscarriage is a far more common problem in middle-aged women than Down's syndrome, and indeed it is one of the major reasons why women in their forties produce fewer babies (the other reasons are that they have less sex and that their cycles are erratic). Yet because miscarriage is usually a private tragedy, and a tragedy which may be followed by the joys of subsequent successful pregnancies, it is less discussed than chromosomal abnormalities.

For reasons we do not understand, older mothers may also be more likely to give birth to smaller babies. Published studies differ in how dramatic they claim this effect to be, but there is certainly evidence that children born to mothers over forty are more likely to fall into arbitrarily defined 'low-birth-weight' categories. Some even suggest that women in their thirties produce smaller babies than women in their twenties. Strangely, the other group of mothers which tends to produce small babies is teenagers, although much of this effect may be explained by those mothers' socio-economic disadvantages. However, some researchers have claimed that the similarity between middle-aged and teenage mothers reflects very real biological concordances between the two groups – for example, in experimental animals, juvenile and older mothers both show abnormal patterns of hormone secretion before and during pregnancy. Another possible cause of reduced weight in babies born to middle-aged mothers is that short intervals between pregnancies are known to reduce birth-weight, and we have already seen that such short intervals are a distinctive feature of today's late-started families.

Apart from these concerns, the news for middle-aged mothers is good – there is no evidence that prematurity is more common in babies born to middle-aged mothers, nor that those babies are more likely to be unresponsive immediately after birth, or to die soon after birth. One study even suggests that outcomes for twins and triplets are better if their mother is middle-aged.

The third potential disadvantage of middle-aged parenthood is commonly ignored, yet it is perhaps the most unexpected and intriguing: advancing paternal age may also adversely affect children's health.

We have all heard stories of men siring children at an advanced age (Charlie Chaplin became a father at seventy-three; the Spanish singer Julio Iglesias's father, appropriately enough a gynaecologist, sired a child when he was eighty-nine) and I suspect that such wrinkly reproductive feats, however uncommon, appeal to the sliver of machismo present in all human males. I think many of us also assume that elderly paternity is without adverse consequences, mainly because it has no obvious effect on the incidence of well-known conditions such as Down's syndrome.

Indeed, unlike maternal age, advancing paternal age has little effect on the incidence of chromosomal abnormalities in children. Admittedly, there may be a very slight effect of paternal age on the occurrence of Down's syndrome, but apart from that tentative link there is no good evidence that older fatherhood causes aneuploidies. The common perception is that sperm gush, short-lived but plucky, from a continually refreshed testicular fount – contrasting sharply with the decades-long wait of the vulnerable egg inside the ovary. This simplistic view of male biology has led to a widely held belief that paternal age is no barrier to conceiving healthy children.

However, recent studies suggest that all may not be well with the ageing male parent. One group of diseases is now generally accepted to be more common in the children of older fathers. These are 'autosomal dominant' genetic diseases, which are caused by damage to just one of the two copies of a single gene which we each carry on our twenty-two pairs of 'non-sex' chromosomes. Although the genetic rules which underlie these diseases are similar, damage to different genes can have strikingly different effects: Marfan syndrome causes limb elongation and heart abnormalities, achondroplasia results in a relatively common form of dwarfism, polyposis coli entails polyps in the large intestine, Apert syndrome leads to malformation of the skull, face, hands and feet, and naevoid basal cell carcinoma syndrome causes the spontaneous formation of tumours. And these are just a few of an increasing list of autosomal dominant conditions which have been linked to paternal age.

While this list may sound alarming, it is important to realize that advanced paternal age does not necessarily account for the majority of cases of all these conditions (although it has been suggested that it may be the primary cause of some of them – achondroplasia, for example). Also, none of these diseases is common – certainly none of them is as common as Down's syndrome, so the risk of any of them affecting an individual pregnancy is small.

Entirely unlike these genetic diseases, with their simple and clear-cut causes, the second major group of diseases which occur more commonly in children of older fathers includes several brain diseases – some with the most impenetrably complex genetics of any known illnesses. For example, an especially intractable target of modern genetic science, schizophrenia, is probably more common in children of older fathers. In fact, some studies suggest that middle-aged fathers may be up to three times more likely than men under twenty-five to have children who eventually develop schizophrenia. Some researchers claim that a quarter of all cases of the disease result from the siring of children by older fathers. However, others disagree and suggest that the effect of paternal age is much smaller. Even so, other brain diseases, including Alzheimer's disease, bipolar disorder and epilepsy, are now also suspected to be partly caused by older fatherhood.

These are complex, subtle diseases of the most complex, subtle organ in the body, so perhaps it is unsurprising that slight effects of paternal age on children's health become apparent in the functioning of the human brain. The sheer complexity of the genetic processes underlying the activity of the brain, and the precision with which its different components must function and interact, make the human brain inherently vulnerable. It is, perhaps, almost too complex for its own good. Even in apparently healthy individuals, paternal age may have negative effects on the brain. People with older fathers score lower in cognitive tests, and some researchers claim that up to two per cent of all the population-wide variation in human intelligence is caused by differences in paternal age. Intriguingly, the effect of maternal age on intelligence may be the opposite, perhaps implying that older women should seek out toy-boys to sire bright children. (There will be more on these Byzantine manoeuvres in the final chapter.)

It now seems that all these cognitive effects may add up to a real problem, because some reports suggest that mortality rates between birth and eighteen are eighty per cent higher in children fathered by men over the age of forty-five. The two main causes of this increased mortality are claimed to be congenital abnormalities, and injuries and accidents. Whether the injuries and accidents result from reduced children's intelligence or reduced paternal watchfulness is unclear.

So although the ageing male reproductive system does not cause chromosomal abnormalities, it is far from perfect. Yet the fact that advancing paternal age leads to very different effects from advancing maternal age should not surprise us, because the cellular basis of male reproduction is so very different to that in females. Women's ovaries may release a few hundred eggs in their lifetime but men often ejaculate more than one hundred million sperm in one go. While eggs wait for decades, waiting to be ovulated, millions of testicular stem cells are continually dividing to produce the hordes of spermatozoa required for male fertility. In female mammals, cells probably divide no more than twenty-two times before becoming an egg cell, whereas in males there is almost no limit to the cell divisions which might precede the formation of a sperm. And the older a man is, the more divisions will have taken place before each sperm is made. It is this frenetic testicular activity which may explain the genetic disadvantages – simple and complex – of older fatherhood. By the time a sperm cell is released from a middle-aged man's testicles, it may be genetically disrupted, its repair mechanisms may be failing, or it may be damaged by our old nemesis from Chapter 2, reactive oxygen species.

Remarkably, there is now evidence that humans evolved mechanisms to alleviate the adverse effects of paternal ageing. We have already seen that older men are no more likely to father children with chromosomal abnormalities. And indeed, men over sixty who still produce viable sperm show little increase in chromosomal abnormality in their testicular cells. In contrast, older men who no longer produce viable sperm *do* exhibit dramatic increases in testicular chromosomal abnormality (although this can of course not be passed on to the next generation). However, we do not know whether sperm production stops because chromosomal damage has occurred, or if cessation of sperm production allows previously active chromosome-repair mechanisms to be switched off, but all the evidence suggests that male reproductive ageing is not simply a process of uncontrolled deterioration. It has checks and balances built into it by millions of years of natural selection. Even middle-age erectile dysfunction has been claimed to have evolved as a mechanism to prevent older men producing children.

The fourth and final potential negative aspect of middle-aged parenthood is postnatal parenting, and the complex psychological effects which parental age may have on children. Is a child's life different if their parents are forty, rather than twenty, years older than them? And does paternal age exert a more variable influence because men can father children at sixty, while women are unlikely to bear children beyond the age of forty-five? Unfortunately, it has been difficult to answer these questions, mainly because it is impossible to carry out experiments on human parenting.

The grown children of middle-aged parents are more likely to complain in retrospect that their mother and father were inattentive or had little emotional or physical energy to invest in them. They may also feel concerned that their parents are likely to become a burden to them sooner than their peers' parents. And older parents simply cannot win this argument: studies show that when their children are not worrying about the future care burden of surviving parents, they are resentful that they may have to live much of their adult life without a father, or that their children may never meet one of their grandparents. Studies suggest that fear of parental (and especially paternal) death can have effects on adolescent psychological development. Also, the youngest children of large and protracted families may consider themselves to be 'tailenders' – unwanted accidents – regardless of their middle-aged parents' reassurance to the contrary. Children born to older parents are also more likely to be only-children, and it has been claimed that they may subsequently blame their doting, financially secure parents for the boredom they experienced due to a lack of siblings. They may blame their 'lone status' entirely on the age at which their parents chose to have children, and may even feel isolated from schoolmates who have younger parents.

However, these children are more likely to result from planned pregnancies and may be more avidly 'longed for' than children of younger parents. And the greater age gap between the generations can bring distinct benefits. In fact, it is older fathers who seem to have the greatest effect, although this probably results from the biological fact that paternal age may vary more than maternal age. For example, older fathers may take on more the role of friendly mentor than the active, involved nappy-changer. Maybe all those ancient changes in the middle-aged brain – the refocusing of social priorities on a few, choice individuals, and the urge to bequeath cultural information – really do make middle-age parenting different.

So parenthood in middle age is a fiendishly complex thing. It does seem to be biologically inferior to young-adulthood parenting, and perhaps this should not surprise us. However, middle-aged parenthood does not seem to be *very* inferior to the youthful version – chromosomal abnormalities become more common with maternal age, but are still not as common as some of the scare stories might make you believe; and many genetic diseases increase in frequency with paternal age, yet none of them is common.

There is a pattern here, and it has important implications for human evolution, as well as any contemporary couple's decision to have babies. The adverse biological consequences of middle-aged parenthood are severe when they occur, but they are infrequent – certainly compared to all the more mundane causes of childhood mortality which have plagued us over the course of human evolution. As a prehistoric couple approached forty, they stood a good chance of producing more healthy children together, and subsequently surviving to rear them to maturity. In a few rare cases a child might be born with a severe genetic or chromosomal disease, but would probably die soon after birth, thus freeing its parents from any further investment in it. It may sound callous to think of ancient human babies in these expendable, almost calculating, terms, but the fact remains that the odds have been stacked in favour of middle-aged parents throughout human history.

17. Is the 'empty-nest syndrome' real?

While some middle-aged people are busy bringing new children into the world, for many, middle age is the time when children leave.

One of the most consistent phenomena of middle age is the departure of children from the family home. In most mammals it is the parents, usually the mother, which actively cause the separation between parents and offspring – who strike the child from the teat, or cast it from the nest, or simply leave it behind to fend for itself. The separation process in humans is unusual in two ways: it does not promptly follow weaning (there are many years of post-lactation nurturing in humans), and also it is often the offspring who mastermind the separation. Human teenagers actively reject their parents by distancing themselves both physically and psychologically. Thus, the empty nest fits our distinctive-abrupt-unique triad of human middle age – distinctive because children usually leave during their parents' middle age, abrupt in that they do it relatively suddenly, and unique because human offspring often leave of their own free will. This urge to achieve independence is built deep into adolescent behaviour, and there is evidence that it is essential for young humans' subsequent well-being as autonomous, happy adults. Thus it is probable that for most of our evolutionary history, young humans drifted from their parents in their mid-teenage years, even if they remained close by within extended family or tribal groups.

At first sight, this unusual child-instigated separation seems to favour the offspring while casting parents adrift. This is especially true when you consider that throughout this book I have argued that the human life-plan is predicated on the tremendous investment we put into our offspring. Calories, care and culture: adult humans are geared up to pump resources into offspring. So what happens to an individual human parent when the need for that investment is gone and the nest is empty?

In developed countries, parents now potentially have more time to spend in the empty-nest phase of their lives. This is partly because people now live longer, but also because patterns of childbearing changed during the twentieth century. Although some women are now starting families later, many women still start – and cease – childbearing in their twenties or thirties, meaning that they are still quite young when their youngest child leaves home. Despite this recent increase in the potential chronological length of the empty-nest 'phase', it is worth asking whether the empty-nest 'syndrome', in its 'classic', pathological form at least, exists at all. Does the departure of children always have negative results? Does it primarily affect women? And is the syndrome unvarying and inevitable?

Medical texts of the 1960s described 'empty-nest syndrome' as a consistent reaction by mothers to the impending or recent departure of teenage children, and stated that this reaction takes two different forms. The 'overt' form involves obvious, expressed anger as well as violent arguments with the fleeing children, whereas the 'latent' form manifests itself as unfocused dissatisfaction and depression. The overt form was claimed to occur in women who were less well educated, had become pregnant earlier in their marriage, had fewer friends, did not work and, most of all, were not married. One confusing assertion was that the overt form tended to occur in women with 'European values', whereas the latent form was more common in women with 'American values'.

While the tone of these early descriptions may seem strange, they raised questions about the empty nest which remain valid today. For example, is it easier to deal with your children's departure if you have a job, friend or spouse you can redirect your attention towards? Are women hit harder than men? Does an individual's experience of the empty nest depend on their pre-existing temperament?

For nuclear families based around a heterosexual couple, the departure of children obviously represents a moment of flux following a period of relative stability. Also, many people believe that 'empty-nest syndrome' is real, and that they have experienced it themselves. They say it involves feelings of sadness and aimlessness, and is often characterized by spending time in departed children's bedrooms and haunts in order to reminisce, or regain a vestige of their previous emotional link with their child. Reports of empty-nest syndrome peak at the start of the academic year, as young people leave home for college or university. The other peak time is at the marriage of a son or daughter, often long after the child has actually left the parental home – a clear sign that empty-nest feelings can be elicited by children's symbolic departure just as much as their physical departure. Some parents even experience pangs of sadness when their youngest child starts school.

But what if all these reports of empty-nest syndrome are just manifestations of simple, run-of-the-mill reactions which anyone would have to a significant realignment of their emotional world? If this were indeed the case, then the empty nest would cease to be a distinct 'syndrome' at all. It would not be a discrete, characteristic part of human psychological development, fixed deep in the mind, waiting to emerge in middle age, but rather a reflection of the simple fact that children often leave home when their parents are middle-aged, and that this *sometimes* makes those parents feel sad.

In fact, research has shown no clear correlation between children leaving home and depression in middle-aged parents, nor with exacerbation of adverse symptoms of the menopause. Indeed, some have reported that the emptying of the nest is associated with increased measures of positive mood and well-being, as well as increased marital satisfaction – although it is unclear whether this is due to the removal of the stress of caring for teenage children, or the boon of increased wealth. Many psychologists now believe there is no simple one-way effect of children's departure on parents' mood. Parents' reactions can range from grief and sadness to relief and liberation. Some experts even describe the empty-nest syndrome as a myth. My own view is that it exists to the extent that it demonstrates just how variable human beings can be in their emotional reactions to the same stimulus.

The empty nest is often claimed to affect mothers more than fathers, but is this true?

Many evolutionary biologists believe that the different

reproductive roles of men and women have inexorably led to a division of labour which remains remarkably similar in most human societies. Whether biological necessity is indeed the cause, it is certainly true that in most cultures women are expected to provide emotional security for their children, whereas men are expected to provide physical security and a secure supply of resources. It has been argued that, as a result of these specializations by the two sexes, a mother's function as a carer wanes during her children's early teenage years, whereas fathers continue to perform their caring role for much longer, especially in large extended-family groups. According to this argument, women lose their caring role in middle age, whereas men retain it into old age, and this is why empty-nest syndrome seems to afflict women more frequently.

However, the links between our evolutionary history and the incidence of post-empty-nest sadness are more complex than that. For one thing, the division of labour between heterosexual partners may not be quite as predictable as is sometimes claimed – men cuddle their children and women gather food, after all, and presumably this has been the case throughout our species' history. Certainly a division of labour seems to be an efficient way of conducting a couple's affairs – it often occurs in homosexual couples, childless heterosexual couples and heterosexual parents whose roles are said to be 'reversed' – but the details of that division of labour can vary between individuals and between cultures. Indeed, some sociologists have challenged the common assertion that men were the main 'breadwinners' over most of human history – instead they view the period between 1870 and 1960 as an atypical phase in western civilization, when a cult of female domesticity temporarily distorted the normal allocation of roles between husbands and wives.

Whatever the truth, the departure of children is a surprisingly complicated thing, and we should not be surprised if fathers and mothers react to it in different ways. First of all, studies have shown that men with children at home are more likely to see themselves as masculine, whereas the presence of children in the home has little effect on women's assessment of their own femininity; thus, one might expect the empty nest to have more effect on fathers' self-image than mothers'. Second, middle-aged parents are also affected by what other people think of them. For example, studies show that people's first impression of a man with a child in tow is that he is more likely to be generous and have high socioeconomic status, whereas women with children are perceived as less ambitious – and presumably, these external perceptions change after middle aged couples' children have left home. Third, there is evidence that mothers and fathers may respond differently to children who have flown the nest. Some psychological studies suggest that fathers are buoyed up by their inherently positive or proactive attitudes to departed children – it is claimed that fathers tend to view their sons as a second, vicarious chance at life, and that they often retain vivifying urges to protect their daughters long after they have left the parental home. Conversely, it is suggested that mothers' attitudes to their absent children are more equivocal, for example that they envy the sexual and educational opportunities available to their daughters.

A further problem with 'empty-nest syndrome' is that it oversimplifies the ways in which families function. It reflects an assumed view of all families as consisting of two lifelong-bonded heterosexual parents, a small number of children conceived at a time of their parents' choosing, and a demographic and economic world in which older relatives and departed children live financially independent lives some distance from this family nucleus. This family structure may have been prevalent in the middle classes in developed countries in the 1960s when 'empty-nest syndrome' was invented, but it has become progressively less common since. More importantly, it is likely to have been less common earlier in human history, too, as well as during that vast stretch of human *pre*-history when we did most of our evolving.

Rather than focusing on the symptom of a recently invented 'syndrome', we could instead ask: when is the nest *meant* to become empty? And do children currently leave the parental home at an 'unnatural' time which makes their departure harder for their parents to bear?

For many modern humans, the empty nest starts sixteen to twenty years after the date when they *decide* to conceive their last child. Thus, the empty nest is now usually determined not by a couple's biology, but by their prior family-composition preferences. This was not the case in the past. For most of human evolution, many adults lived through middle age and beyond, and women conceived children until their fertility waned naturally after the age of forty. Because of this, humans evolved to enter the empty-nest phase some time in their late fifties. Indeed, one could argue that middle age itself could be defined as the period of care for a last-born child of a freely-copulating couple who do not use contraception – perhaps between the ages of forty-two and fifty-eight.

One corollary of calculating the 'natural' age of the empty nest is that today the nest often empties unnaturally early, because of decisions couples make to artificially curtail their fertility. In fact, it could be argued that it is the 'career women' who delay conception until forty who will eventually enter the empty-nest phase in their late fifties – the 'natural' time at which it has happened over most of human history. It is intriguing to speculate whether these late-childbearers are actually the ones who find themselves in an empty nest the age at which they are best evolved to cope with it.

From children's point of view, however, our surprisingly late estimate for the 'natural' time of the empty nest does show that human parenting is inherently flexible. Throughout human history, children have been raised either by young adult parents or by middle-aged parents – it simply depended on their parents' age when they were conceived. In fact, it is an exceptional feature of human biology that children may be raised by parents who are either fully fertile or completely infertile – this simply does not happen in most other animals.

Another aspect of the empty nest which remains unresolved is the extent to which children are 'meant' to absent themselves from their parents as they become independent. We know surprisingly little about the family and social structures which prevailed during most of human evolution, although we often assume that prehistoric humans lived in small nomadic groups, each based on several extended families. We do not know if environmental and cultural forces tended to keep adult children near their parents, or whether they pushed them away. For example, it is likely that inbuilt urges to support genetically related kin kept extended families together, but it is also probable that families dispersed to avoid competition for scarce resources. In addition, anthropologists have suggested that early humans developed a tradition of deliberately swapping young adults with neighbouring tribes to prevent the potentially disastrous inbreeding which can blight small isolated groups of animals.

In developed countries, economic necessity has led to dramatic distortions of the empty-nest phenomenon. Offspring can now move large distances from their parents to pursue their careers, and this may lead to profound emotional separation. This 'career nomadism' became very marked in the last half of the twentieth century, and is greater in families with higher socio-economic status. Its emotional effects can also be exacerbated by the recent tendency of middle-aged women to re-enter the employment market, and thus have less time to travel to their distant children. However, over the last few years this emotional separation has been partially alleviated by the advent of modern electronic communications. Whether Facebook and Skype can recreate the tribe, we must wait and see.

At the other end of the spectrum are offspring who, often for economic reasons, either return to the parental home – 'boomerang children' – or simply do not leave it in the first place. Until recently the main reasons for this phenomenon were unemployment, the financial penury of higher education, or simple inertia. Today the high cost of housing has become a further reason why the nest does not empty on time – in some countries many young adults live with their parents for precisely this reason, and those parents often respond by subsidizing their offspring's first property purchase, or even encouraging marriage so that two young-adult incomes may be pooled to reach the first rung of the property ladder. However, economics alone cannot explain boomerang children, because family structure also plays a role. For example, youngest children are more likely to stay at home than their older siblings, and first-born children are the least likely to do so – maybe young people leave the parental home to escape siblings as much as to escape parents. Also, remarriage of a parent to a new spouse seems to be a strong deterrent to a child thinking of lingering under the parental roof.

In fact, people's reactions to children who stay at home demonstrate that the empty nest is actually an important, beneficial and desirable phase of life. Middle-aged parents often report being embarrassed that their home-dwelling adult children are not fully independent, employed or married. Indeed, research shows that middle-aged parents' well-being is closely linked to their perceptions of their children's independence and success in the outside world. Studies also suggest that parents often feel guilty because they resent the disruption to their lives caused by the constant presence of their lumbering adult offspring. Also, while there is no documented effect of 'unemptied nests' on parental mood, there are reports that they lead to decreased frequency of sex in middle-aged couples. Although we may joke about it, maybe the empty-nest phase really is a time when couples can look forward to once again parading around scantily clad without fear of their children's vocal revulsion.

Another often ignored aspect of the empty-nest phase is that it frequently overlaps with the time when middle-aged people start to worry about their own parents' health. We inhabit an unusual time in human history when old-age enfeeblement no longer rapidly leads to death, and middle-aged people often find themselves caught in the middle – a 'sandwich generation' who must simultaneously care for non-departing children and ailing parents. And it is the 'ailing' which is the important bit, because middle-aged people are usually able to successfully renegotiate relationships with *healthy* elderly parents on a more equal, peer-like footing. In fact, studies show that healthy parents can be a source of great support, whereas unhealthy ones can be a new burden to replace the 'old' burden of children – often leading to career stagnation and financial hardship in the sandwiched middle-aged who must care for them. It is important to realize just how unnatural this situation is, because throughout most of human history unhealthy elderly parents soon departed, and in a more irreversible way than teenage children do.

So, although the empty-nest phase is one of the most consistent features of human middle age – and could even be the one thing which numerically defines what middle age *is* – its effects are not at all consistent. To some it is a liberation, to others a curse. It is not an inbuilt psychological reaction, but simply a normal response to the fact that offspring tend to leave their parents when they are middle-aged. The concept of a 'pathological' empty-nest syndrome was developed at a time and place when society's view of women's roles was unusually focused on rearing children. It was an attempt to medicalize the normal, and as a result it focused on people whose reactions to the empty nest were negative.

It might even be argued that there is only one reason why the emptying of the nest should necessarily be a negative thing, and this relates to the asymmetry in the relationship between parents and children. If a teenager feels nervous or depressed about leaving the family home, then they often feel free to express their feelings to their friends or family. Yet middle-aged parents who experience the same feelings are emotionally isolated – they often believe they cannot discuss their worries with their friends, and certainly not with their children, for fear they will be seen as indulging in overprotective emotional blackmail.

But life is not fair, and indeed parents' insecurities should not stymie their children's first forays into the outside world. Evolutionary theory tells us that young people are, quite simply, more important than older people, so middle-aged empty-nest parents must sometimes bite their tongues. And then, presumably, get on with having more independence, fun and sex.

18. Will you still love me tomorrow?

Many couples wonder what their romantic and sexual lives will be like when they reach middle age. This is a period in life during which, if the couple has had children who have now left home, the relationship will necessarily turn back in on itself, as a unit of many reverts to being a unit of just two. And for any couple – parents or not, heterosexual or homosexual – there are, as we have seen, numerous physical and psychological shifts taking place within each partner as they edge past the midpoint of life. If they are no longer quite the same people, how will they feel about their partner, and how will their partner feel about them? In middle age, long-standing romantic relationships become precariously exposed. With the old youthful urges to fall in love, succeed in life and rear children receding, people can suddenly feel alarmingly dependent on their partner's unpredictable responses to all this change. It is an eternal paradox of human life that although we are an obsessively social species, we can never truly get inside other people's heads. As much as we may want to do it – and in the case of our romantic partners we may want to do it *a lot* – we can never know exactly what another person is thinking, or what they will do. This uncertainty may be exciting in our twenties, but in middle age it can become frighten-ing. How is the person with whom you have constructed your entire adult life going to react now you and they are older, alone, together?

At first sight, our understanding of evolution and natural selection makes this situation appear even more dicey. In humans, as in any mammalian species which forms pair-bonds to rear their offspring, as those offspring grow up, their parents' need to invest in them decreases and the option of ending their partnership and starting afresh becomes less off-putting. Also, economically, increased wealth renders middle-aged couples more able to cope financially if they split up. To make matters worse, this is a time when, in het-erosexual relationships, the differences between the sexes become especially unfair. Middle-aged women worry that they will not be able to create a new life if their relationship dissolves – their attractiveness is perceived to decline more rapidly than men's, they earn less, especially if they have taken a career break to care for children, and their fertility is declining. Men face none of these problems and conse-quently have more romantic options in middle age – including the option of effectively engaging in serial monogamy by switching to a new long-term partner. The perception that men have greater flexibility to move on to new relationships creates a tension, often unspoken, in almost all middle-aged relationships.

The aim of this last chapter is to find out what happens to romantic relationships in middle age. Do uncontrollable forces conspire to make them inherently unstable, or is there a secret glue which holds them together? How many middle-aged people stray from their partner and why? I will also consider whether failed middle-age relationships signify that humans never evolved to be monogamous in the first place. Do we deny our nature by staying together in middle age?

Estimates published by the United Nations indicate that eighty-nine per cent of all humans have married by the age of forty-nine. This is a striking statistic, and it is no doubt a major reason why many research studies have focused solely on married couples – after all, legally recognized heterosexual relationships represent a large, readily identifiable sample group just begging to be studied. Of course, many people are homosexual and many heterosexual people are unmarried, so data from married couples can never give us the whole picture. There is a great deal of research to be done on how 'non-marriage' relationships change during their middle years, but their dynamics may not turn out to be so very different. After all, at present most homosexuals and unmarried heterosexuals inherit their genes, and thus at least some of their behavioural traits, from their married heterosexual parents.

The world of long-term relationships has changed recently in developed countries. More couples now cohabit rather than getting married (more so in Europe than in the US), yet marriage still predominates – most people still get married at some point in their lives. More people are juggling careers and childrearing, and men are playing a greater direct role in childcare. Also, more people are getting divorced, and fewer divorcees are remarrying (although it is likely that this increasing rate of divorce is also reducing the number of silently loveless, abusive or otherwise unsatisfactory marriages). All these trends mean that people are spending a smaller proportion of their lives married.

There are striking differences in rates of marriage in men

and women at different ages. The percentage of men who are married or cohabiting steadily increases throughout adult life, whereas in women it peaks somewhere in middle age – between the ages of forty-five and fifty-five in developed countries. This discrepancy has two main causes: women are likely to outlive their husbands, and women are less likely than men to remarry after separation or bereavement. Because of this, unmarried widowhood is eventually *five* times more common in women than in men. Thus, whatever may be said about men in their youth, they truly are the marrying kind later on, and being unmarried dramatically reduces their life expectancy – an effect not seen in women. We do not have to wait until old age for these trends to become apparent either, because more middle-aged men are married than middle-aged women (eighty-two per cent, as opposed to sixty-nine per cent in one study). This is probably why newspaper articles appear regularly, penned by middle-aged women, bewailing the scarcity of eligible middle-aged men. Obviously these contrasts cannot be explained by differences in longevity, because we have already seen that few people die during middle age. Thus, an appreciable number of middle-aged men must be married to women who are not themselves middle-aged: these women are mostly in their twenties and thirties, and many of them are second wives. As we have already seen, this probably has more to do with men and women's inbuilt age preferences for their partners than any 'mid-life crisis', yet the asymmetry of the sexes in middle age is still striking.

Yet the statistics do not show that middle age is especially plagued by relationship breakdown. Quite simply, most divorces occur during the third and fourth decades of life, which is intriguing, considering that this is the time when the flame of romance and the embers of procreation are supposed to burn most strongly. In comparison, middle age emerges as a relatively stable time for romantic relationships (especially if it is assumed that the figures for middle-age divorce are bolstered by the many couples who divorce having planned to do so many years before, but who decided to stay together until their children had grown up). Middle-aged marriage seems to be surprisingly robust, despite the pressures and fears associated with it. Clearly, there must be forces which shore it up.

Love usually occurs in several phases. Initially it is an entirely within-one-mind phenomenon as one individual falls in love with another. Then comes the exciting stage when both parties undergo a tumultuous mutual psychological realignment – they become naïve about each other, idealize each other, and deny any criticism of the object of their affections. This phase of love is not just psychological, but also involves changes in body chemistry: certain chemicals have recently been shown to become more abundant in the blood when people first fall in love, and they even attain higher levels in people who feel particularly smitten. In the next phase, the relationship slowly becomes calmer and more realistic, and many of the measurable chemical changes of love subside. Presumably this chemical and psychological shift is the time when humans evolved to divert their attention away from their partner and towards the babies their mutual passion had generated. By the time a couple who met in young adulthood reach middle age, this transition from chemically fired passion to refocused caring is long past.

Studies suggest that there are several negative influences at work in failing middle-age relationships, all of which are less important in failing relationships in younger people. Boredom and devitalization is often reported, and events often conspire to make this a time when couples detect a low-point in companionship and affection. Old bones of contention may come to the fore – for example, studies show that sexual dissatisfaction in decades-long marriages is more likely when one partner had many more before-marriage sexual partners than the other. Also, although it is a cliché that middle-aged couples do not communicate with each other, some therapists believe the opposite to be true – that middle-aged partners are so good at subliminal communication that they often negotiate undemanding, loveless, affectionless relationships without ever discussing them verbally.

Women in particular seem to have a hard time with their relationships in middle age. Middle-aged women may openly express their disappointment about their partner for the first time – often criticizing their passivity, their unwillingness to change, their emotional dependency, or their inability to provide. Research suggests that married women's self-esteem is more dependent on the perceived romantic commitment within their marriage, whereas men's self-esteem is more directly related to their sex-life. In unhappy marriages, women are more likely to report mental and physical health problems than their husbands. Also, if they enter relationship counselling, women are more likely to be suffering from negative moods, whereas men are more likely to be defensive. In addition, although evidence suggests that women worry most about their looks when they are young adults (in the twenties for unmarried women and in the thirties for married women), these worries decline only slowly during middle age, and often transiently increase as their daughters start to date, or as they see their own future romantic options narrowing while their husbands' widen.

Men report more marital satisfaction than women. This may of course be genuine – we have already seen that men can be inherently bullish – or it could reflect the possibility that men are less likely to report problems. Some men even report being bewildered by their wife's dissatisfaction with what they themselves view as a perfectly good marriage. Also, research suggests that simply being married is more important to men than the actual 'quality' of that marriage. Apart from the importance of sex mentioned above, another thing which contributes to male marital happiness is a perception that 'orthodox' gender roles are being maintained within the relationship. As we have already seen, couples usually divide their emotional and practical roles asymmetrically, and this is a powerful urge, especially for men. Indeed, so strong is this urge that homosexual couples frequently develop similar asymmetries in their relationships – suggesting that the human drive to partition relationship roles is even more prevalent than the urge to procreate.

Yet middle-aged humans are complicated things, and concentrating on marriages which are failing can give a misleading idea of what is going on. Most marriages, after all, do not fail in middle age. Middle-aged people still seek out the positive in their partners, even if what they seek is different from twenty years earlier – they are more likely to cite characteristics such as tolerance, reciprocity and trustworthiness, for example. To young adults those may sound like dull, conservative traits which simply serve to *protect* you from your partner rather than things which might inspire, haunt and seduce you, but perhaps some of us have had our fill of inspiration, haunting and seduction by the time we reach fifty.

Physical attractiveness remains important in middle age, but other traits come to the fore as well. We all know that intelligence, humour, kindness and creativity are attractive in a partner, and evolutionary biologists believe that we like them because they are signs of genetic health and an ability to care and provide for offspring. However, they have a further important role in human middle age because they survive the deterioration of our visual appearance, and may even subconsciously remind us that the crinkly, saggy, tubby being in front of us really *is* the same person as the pert young thing we once hooked up with. Attraction to these non-deteriorating traits may well hold couples together throughout middle age and beyond.

Biologists have also suggested that, unexpectedly, middle age may be a time when natural selection favours the good guy. When females select a male to father their children, they often choose the mate who appears to have the best genes. However, recent studies have shown that it may make sense for a female to choose a mate who is slightly sub-optimal in looks and physique, if that male shows signs that he is likely to lavish parental care on his offspring. Natural selection is all about producing successful offspring, but in many species, especially humans, this requires not only good genes, but also good paternal care. Thus, middle age may be a time when women who married one of these genetically sub-optimal but parentally excellent males finally win out. According to this theory, these males continue to care for their children as their partner ages, while flashier, more genetically perfect males wander off to find another mate.

It is all very well describing what happens in modern-day marriages and extrapolating these findings to other long-term relationships, but one crucial question hangs over us. Is lifelong monogamy the way we evolved to live our lives? Monogamy may be the dominant form of heterosexual relationship in developed countries today, but it is possible that it is no more than a modern artificial invention. If we were to look back in human history and discover that lifelong monogamy was not the norm, then our entire view of the natural history of modern marriage would have to change. Are men *meant* to leave their partners in middle age and try their luck with new ones? Or, for that matter, are women meant to? Are humans naturally serially monogamous, polygamous, or even promiscuous?

In other words, are the trials of middle-age relationships evidence that we have imposed an essentially unnatural system of breeding on ourselves?

Monogamy is rare in mammals – between three and five per cent of all species are monogamous, including prairie voles, beavers, dik dik (a tiny antelope) and some marmosets and bats for example – and these tend to be species in which offspring receive care from both parents. However, some groups of animals utilize monogamy much more than mammals – approximately ninety per cent of birds are monogamous (serially or lifelong), and once again, in most of these species both parents care for their offspring. Yet among our closest relatives monogamy is uncommon – the great apes are either promiscuous or polygamous – so at some point in the very distant past, it is likely that the ancestors of humans were not monogamous.

Some anthropological studies have suggested that polygamy is still the predominant reproductive system in humans, present in approximately three-quarters of all cultures. However, the fact that polygamy is permitted does not mean that most men employ it, nor that polygamy represents crude male sexual acquisitiveness. Most married men in 'polygamous societies' have one wife, usually as a reflection of their social or economic status. Also, if men do marry additional wives, these are often 'inherited' from deceased male relatives – a process which usually occurs during male middle age. A common pattern in polygamous marriages is that there is a much-loved, dominant, primary wife who lives monogamously with her husband for many years before he marries other wives. One study in Ethiopia showed that first wives of polygamous men seem to do better than wives of monogamous men – they marry younger, implying higher 'marriageability' (wealth, status and beauty, perhaps), and their children are taller, heavier and better fed. Subsequent polygamous wives marry when older and do least well, and it has been suggested that these women are somehow marrying to 'make the best of a bad job'. Thus, polygamy is not as widespread as it may seem, and is not necessarily evidence of selfish subjugation of women by dominant men. (The opposite, 'polyandry', in which one woman marries several men, is uncommon, presumably because it is a very inefficient way for a population to produce children, because most females' fertility is wasted.)

We have conflicting biological evidence regarding the prevalence of polygamy over the course of human history. For example, by tracking the spread of genes on the Y chromosome, it has proved possible to show that some individual ancient men were the direct patrilineal ancestors of huge numbers of individuals – one study suggested that the close relatedness of 0.5 per cent of the entire modern human population is explained by the polygamy of one Central Asian man, presumably Genghis Khan. However, other genetic studies suggest polygamy has been relatively unimportant in human history. Comparing the X chromosome with other chromosomes in the extent to which its genes are shuffled, shows that the ratio of breeding women to breeding men has been between 1.1 and 1.3 during our species' past – suggesting that monogamy has predominated. Also, it should be emphasized that the fact that the ratio is not 1.0 could result not only from our ancestors being occasionally polygamous, but also from their being either serial monogamists, or simply unfaithful.

Studies in other species have shown that certain simple measurements can be an excellent guide to whether animals breed monogamously. For example, size is a good indicator: in polygamous primates, males are often twice as massive as females, whereas in monogamous gibbon species, males and females are the same size. Human males are roughly 1.2 times larger than human females – nearer a gibbon-like 1.0 than a gorilla-esque 2.0 – which implies once again that we lie towards the monogamous end of the spectrum. We get the same result when we examine features of our biology which we know indicate promiscuity in other species, too – males of promiscuous species have large testicles and characteristic sperm morphology to increase the chances of their sperm winning the race to the egg in multiple-mated females. In contrast, human testicles are relatively small, and human sperm look distinctly non-racy.

All this evidence has led some researchers to take the most extreme view and claim that lifelong monogamy is the original mode of reproduction for *Homo sapiens* – and that it is entirely natural for middle-aged men to remain with their menopausal spouses, even though it curtails their own reproductiveness. Certainly, there are good reasons why our species might adopt such a 'bird-like' strategy. For example, parental investment in offspring is exceptionally important in humans, and it could be argued that men can produce more successful offspring by concentrating on caring for the offspring of one woman, rather than siring many children who die, suffer or fail due to paternal neglect. Indeed, there is evidence that men's sexual interest is slanted towards the women who have already borne their children – a study of forager-horticulturalists in Bolivia showed that men are most likely to have 'affairs' when they have fewer children, and this parallels the finding that young adults are more likely to have affairs and get divorced than middle-aged people. (However, we cannot rule out the possibility that middle-aged people have fewer affairs because no one wants to hop into bed with them any more.)

Lifelong monogamy may have other advantages, too. Cooperation between males is essential for the smooth running of human societies, yet this cooperation could be seriously compromised by unrestrained sexual competition between men. Thus, it is argued that men and women who seek monogamous relationships are more likely to be members of successful, peaceful communities. It is not that these people 'choose' to be monogamous, but simply that over the millennia people with genes for monogamous behaviour have thrived. In this context it is notable that in modern sexually liberated societies, most people, heterosexual or otherwise, eventually seek to establish pair-bonds in preference to other romantic and sexual arrangements.

However, many of these theories still do not exclude the possibility that the original human breeding system was one of long-term *serial* monogamy. Indeed, many modern hunter-gatherer societies exhibit just such a system – in which men and women pair up for a decade or more, but then drift apart and find someone else. As they age, the chances of these subsequent pairings producing children decrease, but this does not seem to stop the partner-swapping. In these societies people also sometimes have 'extramarital' sex, especially when they are young.

And many anthropologists now believe that our current system, in which society frowns upon serial monogamy, is actually an artificial creation – a desperate response to our old nemesis: agriculture and the settled life. Researchers have proposed several reasons why farming, settlement, property and inheritance might have led societies to impose lifelong monogamy on their members. First, institutionalized monogamy could have been a simple response to the fact that, after the agricultural revolution, people were forced to live in much closer proximity to each other. In these con-fined, impoverished conditions, it was even more important that men should not compete sexually with each other. Thus, the argument goes, enforcing lifelong monogamy success-fully suppressed this destructive competition – and this may even explain why many societies mete out draconian punish-ments for adultery. A second theory is that following the advent of agriculture, men had more to bequeath to their off-spring than just their genes. They now had property to dole out as well, and having just one wife meant that this property would not be scattered between many descendants. The divi-sion of property is an unavoidable result of having many children, but if each of those children inherits less land than is necessary to feed one person, then siring lots of babies by many mothers becomes completely counterproductive.

I would like to suggest a third reason why farming led to institutionalized monogamy – marriage, in other words –

and unlike many previous theories it also takes account of women's attempts to increase their children's survival and success. Earlier we saw how the coming of agriculture allowed humans to produce more babies than at any time in their previous history. This, of course, is why almost all humans in the world now live in societies based on agriculture – despite its terrible effects, it is a system which *works*. However, a couple who tend a plot of land are constrained in their ability to produce healthy children by the productivity of that plot. If, after siring some children, a middle-aged farmer decides to leave his wife, he faces an awful choice. He could leave her the land (perhaps unlikely) so she can support his existing children, or he could retain the land and start over with a new woman, leaving his previous children to starve. The one option not open to him is to give half the land to his ex and keep half to feed himself, his new wife and any children they may produce. There simply is not enough productive land to feed them all (and if there is, it will soon be subdivided between so many children that his sons or grandsons will eventually face the same dilemma). Thus, middle-aged men cannot leave, middle-aged women get some security, and society imposes lifelong monogamy to formalize an arrangement that is, effectively, already unavoidable.

Thus, a concept has emerged of prehistoric humans as having a choice of monogamies – serial and lifelong – with a few affairs (especially youthful) thrown in. This slightly vague natural breeding system could explain why the genetic and biological data implied that we are an *almost*-monogamous and *almost*-non-promiscuous species, as well as why most people voluntarily seek to be one half of a couple. Yet only later was this natural breeding system

supplanted by a farming-driven need to be lifelong monogamists, whether we liked it or not.

And it is this conflict between our pre-agricultural nature
and our post-agricultural culture which causes so many
problems today. It is individuals' self-evident ability to quit
lifelong monogamy and switch to the serial kind which
looms menacingly over middle-aged people's romantic relationships.

So how common is infidelity in middle age, and is it necessarily a forerunner to separation?

As you would expect, the figures again focus on marriage,
rather than non-married long-term relationships. Some
researchers claim that marital infidelity is so rare that it is
irrelevant. The incidence of infidelity is indeed low in many
countries – perhaps as few as 1.5 per cent of married people
have affairs in any one year – but over the decades these small
numbers add up. It is estimated that between forty and sixty
per cent of married people in developed countries have at
least one affair at some point. Of course, people are loath to
admit to affairs when completing researchers' questionnaires,
and it is also likely that many affairs are so unsatisfying or
traumatic that they are short-lived, not repeated and, indeed,
unimportant – *statistically*, at least.

Yet it is still intriguing to investigate the evolutionary
forces which drive middle-aged men and women to have
affairs. For ancient hunter-gatherer men the potential gains
of fathering a child by an extra-dyadic relationship were considerable (I will call pre-agricultural monogamous pairings
'dyads' because I do not know if they were 'marriages' as we
know them). These men could, after all, have an extra baby
for very little investment, and their infidelity would almost

certainly never be discovered. They could also exploit an unusual feature of human sex, which is that it usually takes place in private – 'private' sex is rare in other species, and it has been claimed that the human form has evolved from the 'covert' sex in which other great apes indulge when they do not want to be detected by their paramour's usual mates. The all-win-no-lose nature of ancient male infidelity may explain why men who have affairs often report acting 'blindly' and instinctively, and can seem staggeringly naïve in failing to consider the consequences of their actions. Perhaps twelve thousand years ago, there *were* no consequences.

A tendency for infidelity does seem to be built into the human male population, although some men have it worse than others. Studies suggest that while some men engage in short-term relationships because of some form of emotional incompetence, in others the tendency is associated with beneficial character traits such as confidence and high self-esteem (this has not been found to be the case in women in serial short-term relationships). Also, recent physiological studies suggest that genes involved in brain processes related to pair-bonding and mate-guarding in a wide variety of species also play a role in deciding whether men have affairs. For example, the release of two chemicals, oxytocin and arginine vasopressin, has been shown to be crucial in the establishment and maintenance of monogamous pair-bonds in some vole species. Differences in the oxytocin-arginine vasopressin system probably explain why some vole species are monogamous while others are not, and similar mechanisms may underlie social bonding in many backboned animals. And intriguingly, men with certain forms of a gene which is used to make a protein which binds to arginine vasopressin are less likely to be married, and if they do marry, their partners tend to have low survey scores of marital satisfaction and perceived relationship strength. In addition, an allied neural system involving the chemical dopamine has also been implicated in infidelity – men with certain variants of a gene which makes a dopamine-binding protein are more likely to report having one-night stands.

The reasons women have affairs are, if anything, even more complex. Because they carry the burden of pregnancy and lactation, hunter-gatherer women would need compelling reasons to have extra-dyadic sex. In many primate species, females engage with multiple sexual partners to achieve various aims: to increase their usual mate's protectiveness, to acquire resources from multiple males, or as a prelude to leaving their original mate for another, better one. And middle-aged human women face a particularly thorny choice, because their fertility is uniquely time-limited. It has been suggested that in common with other female primates, many middle-aged women have affairs as part of a last-minute 'gene-grab': an attempt to conceive a final child by a particularly genetically desirable male, with the intention of returning to their usual partner so he can help care for it. Thus women, for whom conception is such a *committing* step, are no less driven to have affairs – it is just that the stakes are higher for them.

So there are very good evolutionary reasons why middle-aged women should have affairs, and this could explain why middle-aged women are only slightly less likely to have affairs than middle-aged men (studies suggest between ten and thirty per cent less likely). And extra-dyadic sex does have its hedonistic temptations for women, too – women are more likely to achieve orgasm when having sex during an affair than within marriage.

For most of human prehistory, humans did not wear much, had sex in private, and did not have recourse to paternity tests. Is it any wonder that middle-aged people sometimes have improper thoughts?

Speaking of impropriety, there is now a great deal of evidence to suggest that modern-day middle-aged people can be a romantically and sexually risqué bunch. Humans flirt for all sorts of reasons – to attract long- or short-term mates, to relieve boredom, to provoke sexual desire, to reaffirm their own self-image, to present themselves as sexually confident, or to stimulate vigilance in their partner – and in middle age, flirting can become more overt than in young adulthood. Those adolescent days of tentative, protracted courtship are over, and by middle age people are often keen to cut to the chase. However, studies suggest that middle-age flirting is often less about attracting new partners than manipulating existing ones. Sexual jealousy is a phenomenon which is receiving increasing attention from evolutionary biologists. We now think of it as a beneficial trait, which prevents many bad things happening: losing mates, losing social status, relinquishing resources to other women's offspring, or unknowingly rearing other men's children. All of these spurs to jealousy are still potential threats by the time humans reach middle age, and this is a time when our capacity to retain mates by non-jealous means – by being beautiful or fertile, in other words – is failing. Paradoxically, sexual jealousy can become stronger and more relevant in middle age, even though our partners are becoming less attractive to others.

There is now good evidence that this middle-aged sexual ferment leads people into behaviours they might have frowned upon twenty years earlier. Indeed, many middle-aged parents indulge in the very 'irresponsible' activities about which they repeatedly warn their teenage children. In one UK study, one-fifth of people between the ages of forty-five and fifty-four reported having unprotected sex with someone who was not their long-term partner. One US study suggested that fifty-one per cent of pregnancies in women over forty are unplanned, while in the UK the percentage of pregnancies in women aged between forty and forty-four which end in termination is the same as in girls under the age of sixteen. There may be simple reasons for this: surveys indicate that middle-aged women underestimate the risk of getting pregnant; the media constantly overemphasize the prevalence of infertility in older women, and healthcare systems rarely focus on contraception in middle age. However, I would argue that the main causes of these 'irresponsible' behaviours are the drives which were built deep into the human brain because sexual 'irresponsibility' can be evolutionarily beneficial in middle age. For example, rates of unplanned middle-aged parenthood could be partly explained by women's subconscious 'gene-grabs', or men's 'covert' impregnations.

For similar reasons, the incidence of sexually transmitted infections is surprisingly high in middle-aged people. In the UK, eight per cent of new HIV diagnoses occur in the over-fifties, and infection rates in this age group are increasing faster than in any other – they more than doubled between 2000 and 2007. It is sometimes claimed that these findings may be explained by middle-aged people re-entering the world of dating following divorce or death of a partner, but we have already seen that both divorce and death are less common during middle age than young adulthood. Instead, research shows that middle-aged people consistently underestimate their risk of catching STIs, and once again I would suggest that this reflects middle-aged people's inbuilt urges to throw caution to the wind as they reassess their romantic and sexual options.

One behaviour which may have a dramatic effect on the incidence of STIs is swinging – politely defined in one journal article as 'consensual mutual involvement in extra-dyadic sex'. Swinging is an unusual sexual behaviour because it seems superficially to run counter to our understanding of how sexual and romantic relationships work. However, it is now thought that swinging does not involve a complete suppression of couples' sexual jealousy, but rather a refocusing of that jealousy to heighten sexual arousal. Whatever the psychological lure of swinging, anecdotal evidence suggests that it is moderately common in middle age and leads to a disproportionate number of admissions to, and positive diagnoses by, STI clinics.

So the apparent stability of middle-aged relationships conceals a cauldron of bubbling biological urges, psychological dilemmas and evolutionary drives. After the advent of agriculture, humans invented the institution of lifelong monogamy to help us cope with these malign forces as we wallowed in the close-packed, pressurized stench of settled life. Yet still today, the ancient fundamental biological differences between men and women continue to drive a wedge between us, more in middle age than at any other time in our lives. As we enter middle age, as the developmental 'clock of life' ticks unrelentingly into its fifth and sixth decades, men and women become ever more different – their income, their involvement with children, their fertility, their perceived attractiveness. And the most destructive differences of all are those pent-up, blind compulsions we have acquired over the last few million years – compulsions to wrest from life every last drop of reproductive success we can get, regardless of who gets hurt.

Yet the evidence shows that despite all these conflicting urges, middle-aged relationships are surprisingly robust, if somewhat fruity. The clichés do not hold as true as we might think – men are not always trying to dump their wives for a younger model; infidelity is often not driven by an urge to leave a long-term partner; affairs are rarely the sole cause of divorce; and extramarital affairs are often less emotionally satisfying than marriage itself. Middle-age sex and romance do not last because they are tranquil or boring, but rather because they represent the most complex commitment two humans can make to each other. After all, by the time we reach middle age, we have finally grown up.

Conclusion

The View from the Summit

> To you is given a body more graceful than other animals, to you power of apt and various movements, to you most sharp and delicate senses, to you wit, reason, memory like an immortal god.
>
> Leon Battista Alberti, *Della tranquillitá dell'animo*, 1441

So what have we learnt from our journey through the third score of our three-score years and ten?

Middle age is often thought of as a dull, transitional stage of life – a shade of grey between the light of youth and the gloom of age – but our zoological approach to human biology has shown us that it is much, much more than that. We have seen how the exceptionally strange human life-plan evolved over millions of years, as the digital code in our genes was reprogrammed to change the shape of our lives. And middle age has been a crucial element in this process, because prehistoric people's impressive longevity gave natural selection ample opportunity to mould the fifth and sixth decades of life. Humans live in a high-pressure, energy-intensive information economy quite unlike any other animal, and the extreme power of our brains, and the extreme lengths to which we go to rear children with such brains, have led us to develop a phase of life simply not seen in other creatures. Other animals go through a middle period of adulthood, but they never experience anything remotely like our wonderful 'middle age'. As Alberti wrote, we humans have 'wit, reason, memory like an immortal god', and without middle-aged people that could never have happened.

In this book, human middle age has emerged not as a cultural invention of the last few decades, but instead as a specific biological entity built into us in the hundreds of thousands of years before we planted our first crop. Again and again we have seen that middle age is far more than a vague, gradual decline into senescence. It involves too many changes which are *distinctive*, *abrupt* and *unique* for that.

But this still does not mean we have come up with a simple definition of middle age. In fact, we have instead accumulated a list of almost-definitions – one for each of the eighteen chapters of this book. So middle age, variously, is:

- the fifth and sixth decades of life
- a balance between the 'clock of life' and the 'clock of death'
- a unique stage of human post-reproductive life when natural selection somehow still acts on us
- when our role changes from procreation and nurturing to provision and cultural perpetuation
- when the 'soma' really starts to look disposable
- the time when humans are phenomenally energy-efficient
- the cognitive peak of the most intelligent being in the known universe
- when the nature, meaning, value and urgency of perceived time all change

- when lifetime achievement changes from youthful hope to manifest, unavoidable fact
- our perfect balance between cognition and emotion
- our great time of mental stability
- the time when we differ from each other the most
- when sex is redefined as fertility wanes
- the time when looking after existing children becomes more important than conceiving more
- when men's reproductive status is most unlike that of women
- a time when we are still, whatever anyone may tell you, able to produce healthy babies
- the time it takes for the last-born child of a freely-copulating, non-contracepting, fertile couple to grow up
- when romantic pairings are most stable, despite everything which conspires against them

And I think that this list, hard-won in the pages you currently hold in your left hand, is the best definition we could have. I have no glib little definition of middle age, no soundbite, and it would belittle middle age if I did. All this constellation of attributes is needed to define it: none of them can do it alone, but in combination they do. Time is precious in each individual's tenure on earth and there simply are not enough decades for all these things to have their own separate phase of life, so instead we jumble them all together into twenty-or-so years. This pressure of time is the reason why we have eighteen overlapping explanations of what middle age is – eighteen *positive* things which allowed our species to become human.

This, as you cannot have failed to notice, is not a self-help book. As I write this conclusion I am forty-two – just starting to live my own middle age – so I am hardly in a position to tell you how you should be living yours. I have no arcane insights to give you at the end of all this. Yes, I am sure it would be a good idea for you to watch your weight, avoid tobacco, keep mentally and physically active, and not worry *too* much about the future. But that, I am afraid, is it. Middle age involves a complete personal reappraisal of the body, the mind and the self – your relevance, role, desires and direction. And that is the sort of thing everyone must work out for themselves. All I have tried to do is explain why so many of us forty- and fifty-somethings do the weird things we do.

I hope that by the end of this book – our new story of middle age – you do not feel powerless, directed in your every thought and action by the bullying hand of evolution. I admit that sometimes I may ascribe a great deal of influence to our genes, our biology and our evolutionary history, but there is good reason for this. So much of the information needed to function as human beings comes to us in our genes, that a very large part of each individual's nature *must* be defined by them. However, it is still important to realize that there is one way in which humans are different – a way in which they can escape evolution. We all, apparently, have free will. Each individual can, quite simply, *choose* what they do.

So fight against your evolutionary inheritance if you want to. It will not make it go away, but it will not do you too much harm, either. Try to retain a youthful appearance, try to have a late baby, try to do some of the forbidden things you wish you had done in your youth. We are lucky to be alive at the best time in human history to be middle-aged – a time when forty-year-olds can be fairly confident of reaching sixty in fine fettle. We each have the freedom, the time and the wit to do with middle age what we please. Just make the most of what you have, and remember – immortal gods do not spend their middle years in uncontrolled deterioration. They spend them in delicious, productive flux.

ABOUT THE AUTHOR

David Bainbridge is the Clinical Veterinary Anatomist at Cambridge University, and fellow and admissions tutor in the arts and humanities at St Catharine's College. He trained as a veterinary surgeon and has carried out research into pregnancy in man and animals at the Institute of Zoology at Regent's Park, the Royal Veterinary College, and Cornell, Sydney and Oxford Universities. His previous popular science books take a zoological approach to human biology, including pregnancy (*A Visitor Within*, 2000), genes and sexuality (*The X in Sex*, 2003), the brain (*Beyond the Zonules of Zinn*, 2008) and adolescence (*Teenagers: A Natural History*, 2009). He lives in Suffolk with his wife and three children. He was forty when he started to write this book, has a belly, eschews reading glasses, drives a sports car and seems not to have heard of clichés.

ACKNOWLEDGEMENTS

As always I would like to thank my agent, Peter Tallack, and my editor at Portobello, Laura Barber, for moderating my blundering course through the mysterious world of publishing. In addition, several academics have given me ideas which have eventually borne fruit, but I would particularly like to thank Jay Stock who pointed me in the direction of some fascinating literature about human prehistory, and Gavin Jarvis who successfully convinced me that collagen is, indeed, fascinating. And of course I would like to thank my family and friends, an increasing number of whom seem to be younger than me. As for my writing environment, I am grateful to Keith Jarrett, Aaron Copland and Massive Attack.

SELECTED BIBLIOGRAPHY

Abbott, R.A., Croudace, T.J., Ploubidis, G.B., Kuh, D., Richards, M. and Huppert, F.A. (2008). The relationship between early personality and midlife psychological well-being: evidence from a UK birth cohort study. *Social Psychiatry and Psychiatric Epidemiology* 43, 679–87.

Allen, J.S., Bruss, J. and Damasio, H. (2005). The aging brain: the cognitive reserve hypothesis and hominid evolution. *American Journal of Human Biology* 17, 673–89.

Allman, J., Hakeem, A. and Watson, K. (2002). Two phylogenetic specializations in the human brain. *Neuroscientist* 8, 335–46.

Alterovitz, S.S. and Mendelsohn, G.A. (2009). Partner preferences across the life span: online dating by older adults. *Psychology and Aging* 24, 513–17.

Apperloo, M.J., Van Der Stege, J.G., Hoek, A. and Weijmar Schultz, W.C. (2003). In the mood for sex: the value of androgens. *Journal of Sex and Marital Therapy* 29, 87–102.

Arck, P.C., Overall, R., Spatz, K., Liezman, C., Handjiski, B., Klapp, B.F. and Birch-Machin, M.A. (2006). Towards a 'free radical theory of graying': melanocyte apoptosis in the aging human hair follicle is an indicator of oxidative stress induced tissue damage. *FASEB Journal* 20, 1567–9.

Armelagos, G.J. (2000). Emerging disease in the third epidemiological transition. In Mascie-Taylor, N., Peters, J. and McGarvey, S.T., eds. *The Changing Face of Disease.* Boca Raton, Florida: CRC Press.

Atlantis, E. and Ball, K. (2008). Association between weight perception and psychological distress. *International Journal of Obesity* 32, 715–21.

Ayoola, A.B., Nettleman, M. and Brewer, J. (2007). Reasons for unprotected intercourse in adult women. *Journal of Women's Health* 16, 302–10.

Aytaç, I.A., Araujo, A.B., Johannes, C.B., Kleinman, K.P. and McKinlay, J.B. (2000). Socioeconomic factors and incidence of erectile dysfunction: findings of the longitudinal Massachussetts Male Aging Study. *Social Science & Medicine* 51, 771.

Bainbridge, D.R.J. (2000). *A Visitor Within: The Science of Pregnancy*. London: Weidenfeld and Nicolson.

Bainbridge, D.R.J. (2003). *The X in Sex: How the X Chromosome Controls our Lives*. Cambridge, Massachusetts: Harvard University Press.

Bainbridge, D.R.J. (2008). *Beyond the Zonules of Zinn: A Fantastic Journey Through Your Brain*. Cambridge, Massachusetts: Harvard University Press.

Bainbridge, D.R.J. (2009). *Teenagers: A Natural History*. London: Portobello.

Baltes, P.B. (1997). On the incomplete architecture of human ontogeny. Selection, optimization, and compensation as foundation of developmental theory. *American Psychologist* 52, 366–80.

Barker, D.J., Winter, P.D., Osmond, C., Margetts, B. and Simmonds, S.J. (1989). Weight in infancy and death from ischaemic heart disease. *Lancet* 2, 577–80.

Barrickman, N.L., Bastian, M.L., Isler, K. and van Schaik, C.P. (2008). Life history costs and benefits of encephalization: a comparative test using data from long-term studies of primates in the wild. *Journal of Human Evolution* 54, 568–90.

Bellino, F.L. and Wise, P.M. (2003). Nonhuman primate models of menopause workshop. *Biology of Reproduction* 68, 10–18.

Bellisari, A. (2008). Evolutionary origins of obesity. *Obesity Reviews* 9, 165–80.

Berkowitz, G.S., Skovron, M.L., Lapinski, R.H. and Berkowitz, R.L. (1990). Delayed childbearing and the outcome of pregnancy.

New England Journal of Medicine 322, 659–64.

Birditt, K.S. and Fingerman, K.L. (2003). Age and gender differences in adults' descriptions of emotional reactions to interpersonal problems. *The Journals of Gerontology. Series B, Psychological Sciences and Social Sciences* 58, 237–45.

Birley, H. and Renton, A. (1999). The evolution of monogamy in humans. *Sexually Transmitted Infections* 75, 126.

Blanchard-Fields, F (2009). Flexible and adaptive socio-emotional problem solving in adult development and aging. *Restorative Neurology and Neuroscience* 27, 539–50.

Blanchflower, D.G. and Oswald, A.J. (2008). Is well-being U-shaped over the life cycle? *Social Science and Medicine* 66, 1733–49.

Blickstein, I. (2003). Motherhood at or beyond the edge of reproductive age. *International Journal of Fertility and Women's Medicine* 48, 17–24.

Bluming, A.Z. and Tavris, C. (2009). Hormone replacement therapy: real concerns and false alarms. *Cancer Journal* 15, 93–104.

Blurton Jones, N.G., Hawkes, K. and O'Connell, J.F. (2002). Antiquity of postreproductive life: are there modern impacts on hunter-gatherer postreproductive life spans? *American Journal of Human Biology* 14, 184–205.

Bogin, B. (2009). Childhood, adolescence, and longevity: A multi-level model of the evolution of reserve capacity in human life history. *American Journal of Human Biology* 21, 567–77.

Bonsall, M.B. (2006). Longevity and ageing: appraising the evolutionary consequences of growing old. *Philosophical Transactions of the Royal Society of London B: Biological Sciences* 361, 119–35.

Booth, A. and Edwards, J.N. (1992). Why remarriages are more unstable. *Journal of Family Issues* 13, 179–94.

Borg, M.O. (1989). The income–fertility relationship: effect of the net price of a child. *Demography* 26, 301–10.

Borod, J.C. et al. (2004). Changes in posed facial expression of emotion across the adult life span. *Experimental Aging Research* 30, 305–31.

Boserup, E. (1965). *The Conditions of Agricultural Growth*. Chicago, Illinois: Aldine.

Bowles, J.T. (1998). The evolution of aging: a new approach to an old problem of biology. *Medical Hypotheses* 51, 179–221.

Braver, T.S. and Barch, D.M. (2002). A theory of cognitive control, aging cognition, and neuromodulation. *Neuroscience and Biobehavioral Reviews* 26, 809–17.

Bremner, J.D., Vythilingam, M., Vermetten, E., Vaccarino, V. and Charney, D.S. (2004). Deficits in hippocampal and anterior cingulate functioning during verbal declarative memory encoding in midlife major depression. *American Journal of Psychiatry* 161, 637–45.

Brewis, A. and Meyer, M. (2005). Marital coitus across the life course. *Journal of Biosocial Science* 37, 499–517.

Brim, O.G. (1976). Theories of the midlife crisis. *Counseling Psychologist* 6, 2–9.

Brim, O.G., Ryff, C.D and Kessler, R.C., eds. (2004). *How Healthy are We?* Chicago, Illinois: Chicago University Press.

Britton, A., Singh-Manoux, A. and Marmot, M. (2004). Alcohol consumption and cognitive function in the Whitehall II Study. *J* 160, 240–7.

Bromberger, J.T., Kravitz, H.M., Wei, H.L., Brown, C., Youk, A.O., Cordal, A., Powell, L.H. and Matthews, K.A. (2005). History of depression and women's current health and functioning during midlife. *General Hospital Psychiatry* 27, 200–8.

Brubaker, T. (1983). *Family Relationships in Later Life.* Thousand Oaks, California: Sage.

Bukovsky, A., Caudle, M.R., Svetlikova, M., Wimalasena, J., Ayala, M.E. and Dominguez, R. (2005). Oogenesis in adult mammals, including humans: a review. *Endocrine* 26, 301–16.

Burkart, J.M. and van Schaik, C.P. (2009). Cognitive consequences of cooperative breeding in primates? *Animal Cognition* 13, 1–19.

Buss, D.M. (2002). Human mate guarding. *Neuro Endocrinology Letters* 23 Supplement 4, 23–9.

Buss, D.M., Shackelford, T.K. and LeBlanc, G.J. (2000). Number of children desired and preferred spousal age difference: context-specific mate preference patterns across 37 cultures. *Evolution and Human Behaviour* 21, 323–31.

Cabeza, R. (2001). Cognitive neuroscience of aging: contributions of functional neuroimaging. *Scandinavian Journal of Psychology*, 42, 277–86.

Cabeza, R., Anderson, N.D., Locantore, J.K. and McIntosh, A.R. (2002). Aging gracefully: compensatory brain activity in high-performing older adults. *NeuroImage* 17, 1394–1402.

Callaghan, T.M. and Wilhelm, K.-P. (2008). A review of ageing and an examination of clinical methods in the assessment of ageing skin. Part I: Cellular and molecular perspectives on skin ageing. *International Journal of Cosmetic Science* 30, 313–22.

Cant, M.A. and Johnstone, R.A. (2008). Reproductive conflict and the separation of reproductive generations in humans. *Proceedings of the National Academy of Sciences USA* 105, 5332–6.

Carnoy, M. and Carnoy, D. (1995). *Fathers of a Certain Age*. Minneapolis, Minnesota: Fairview Press.

Carstensen, L.L. (1992). Motivation for social contact across the life span: a theory of socioemotional selectivity. *Nebraska Symposium on Motivation* 40, 209–54.

Carver, C.S. (2000). On the continuous calibration of happiness. *American Journal of Mental Retardation* 105, 336–41.

Caspari, R. and Lee, S.H. (2004). Older age becomes common late in human evolution. *Proceedings of the National Academy of Sciences USA* 101, 10895–900.

Caspari, R. and Lee, S.H. (2006). Is human longevity a consequence of cultural change or modern biology? *American Journal of Physical Anthropology* 129, 512–17.

Charles, S.T. and Carstensen, L.L. (2008). Unpleasant situations elicit different emotional responses in younger and older adults. *Psychology and Aging* 23, 495–504.

Charlesworth, B. (1993). Evolutionary mechanisms of senescence. *Genetica* 91, 11–19.

Charmantier, A., Perrins, C., McCleery, R.H. and Sheldon, B.C. (2006). Quantitative genetics of age at reproduction in wild swans: support for antagonistic pleiotropy models of senescence. *Proceedings of the National Academy of Sciences USA* 103, 6587–92.

Childe, V.G. (1951). *Man Makes Himself*. New York: Mentor.

Clark, A.E. and Oswald, A.J. (2003). How much do external factors affect wellbeing? A way to use 'happiness economics' to decide. *The Psychologist* 16, 140–1.

Clark, A.E., Oswald, A.J. and Warr, P. (1996). Is job satisfaction U-shaped in age? *Journal of Occupational and Organizational Psychology* 69, 57–81.

Coelho, M., Ferreira, J.J., Dias, B., Sampaio, C., Pavão Martins, I. and Castro-Caldas, A. (2004). Assessment of time perception: the effect of aging. *Journal of the International Neuropsychological Society* 10, 332–41.

Cohen, A.A. (2004). Female post-reproductive lifespan: a general mammalian trait. *Biological Reviews of the Cambridge Philosophical Society* 79, 733–50.

Cohen, M.N. and Armelagos, G.J., eds. (1984). *Paleopathology at the origins of agriculture*. Orlando, Florida: Academic Press.

Coleman, S.W., Patricelli, G.L. and Borgia, G. (2004). Variable female preferences drive complex male displays. *Nature* 428, 742–5.

Cornelis, I., Van Hiel, A., Roets, A. and Kossowska, M. (2009). Age differences in conservatism: evidence on the mediating effects of personality and cognitive style. *Journal of Personality* 77, 51–87.

Costa, R.M. and Brody, S. (2007). Women's relationship quality is associated with specifically penile-vaginal intercourse orgasm and frequency. *Journal of Sex and Marital Therapy* 33, 319–27.

Cotar, C., McNamara, J.M., Collins, E.J. and Houston, A.I. (2008). Should females prefer to mate with low-quality males? *Journal of Theoretical Biology* 254, 561–7.

Craik, F.I. and Bialystok, E. (2006). Cognition through the lifespan: mechanisms of change. *Trends in Cognitive Sciences* 10, 131–8.

Crews, D.E. and Garruto, R.M. (1994). *Biological Anthropology and Aging*. New York: Oxford University Press.

Crews, D.E. and Gerber, L.M. (2003). Reconstructing life history of hominids and humans. *Collegium Antropologicum* 27, 7–22.

Cutler, R.G. (1975). Evolution of human longevity and the genetic complexity governing aging rate. *Proceedings of the National Academy of Sciences USA* 81, 7627–31.

Cyrus Chu, C.Y. and Lee, R.D. (2006). The co-evolution of intergenerational transfers and longevity: an optimal life history approach. *Theoretical Population Biology* 69, 193–201.

Dakouane, M., Bicchieray, L., Bergere, M., Albert, M., Vialard, F. and Selva, J. (2005). A histomorphometric and cytogenetic study of testis from men 29–102 years old. *Fertility and Sterility* 83, 923–8.

Davidson, R.J. (2004). Well-being and affective style: neural substrates and biobehavioural correlates. *Philosophical Transactions of the Royal Society of London B: Biological Sciences* 359, 1395–441.

de Rooij, S.R., Schene, A.H., Phillips, D.I. and Roseboom, T.J. (2010). Depression and anxiety: Associations with biological and perceived stress reactivity to a psychological stress protocol in a middle-aged population. *Psychoneuroendocrinology* 35, 866–77.

Deary, I.J., Allerhand, M. and Der, G. (2009). Smarter in middle age, faster in old age: a cross-lagged panel analysis of reaction time and cognitive ability over 13 years in the West of Scotland Twenty-07 Study. *Psychology and Aging* 24, 40–7.

Deeley, Q. (2008). Changes in male brain responses to emotional faces from adolescence to middle age. *NeuroImage* 40, 389–97.

Demerath, E.W., Cameron, N., Gillman, M.W., Towne, B., Siervogel, R.M. (2004). Telomeres and telomerase in the fetal origins of cardiovascular disease: a review. *Human Biology* 76, 127–46.

Dennerstein, L., Dudley, E. and Guthrie, J. (2002). Empty nest or revolving door? A prospective study of women's quality of life in midlife during the phase of children leaving and re-entering the home. *Psychological Medicine* 32, 545–50.

Desta, B. (1994). Ethiopian traditional herbal drugs. Part III: Antifertility activity of 70 medicinal plants. *Journal of Ethnopharmacology* 44, 199–209.

Deykin, E.Y., Jacobson, S., Klerman, G. and Solomon, M. (1966). The empty nest: psychosocial aspects of conflict between depressed women and their grown children. *American Journal of Psychiatry* 122, 1422–6.

Doshi, J.A., Cen, L. and Polsky, D. (2008). Depression and retirement in late middle-aged U.S. workers. *Health Services Research* 43, 693–713.

Downs, J.L. and Wise, P.M. (2009). The role of the brain in female reproductive aging. *Molecular and Cellular Endocrinology* 299, 32–8.

Draaisma, D. (2004). *Why Life Speeds Up As You Get Older*. Cambridge: Cambridge University Press.

Drefahl, S. (2010). How does the age gap between partners affect their survival? *Demography* 47, 313–26.

Duetsch, H. (1945). *The Psychology of Women: A Psychoanalytical Interpretation*. New York, New York: Grune and Stratton.

Dukers-Muijrers, N.H., Niekamp, A.M., Brouwers, E.E. and Hoebe, C.J. (2010). Older and swinging; need to identify hidden and emerging risk groups at STI clinics. *Sexually Transmitted Infections* 86, 315–17.

Eagleman, D.M. (2008). Human time perception and its illusions. *Current Opinion in Neurobiology* 18, 131–6.

Earle, J.R., Smith, M.H., Harris, C.T., Longino, C.F. (1998). Women, marital status, and symptoms of depression in a midlife national sample. *Journal of Women & Aging* 10, 41–57.

Ecob, R., Sutton, G., Rudnicka, A., Smith, P., Power, C., Strachan, D. and Davis, A. (2008). Is the relation of social class to change in hearing threshold levels from childhood to middle age explained by noise, smoking, and drinking behaviour? *International Journal of Audiology* 47, 100–8.

Elovainio, M. and others (2009). Physical and cognitive function in midlife: reciprocal effects? A 5-year follow-up of the Whitehall II study. *Journal of Epidemiology and Community Health* 63, 468–73.

Eskes, T. and Haanen, C. (2007). Why do women live longer than men? *European Journal of Obstetrics, Gynecology and Reproductive Biology* 133, 126–33.

Fahrenberg, B. (1986). Coping with the empty nest situation as a developmental task for the aging female – an analysis of the literature. *Zeitschrift für Gerontologie* 19, 323–5.

Farrell, M.P. and Rosenberg, S.D. (1981). *Men at Midlife*. Dover, Massachusetts: Auburn House.

Fedigan, L.M. and Pavelka, M.S. (1994). The physical anthropology of menopause. In Hening, A. and Chang, L., eds. *Strength in*

Diversity: A Reader in Physical Anthropology. Toronto: Canadian Scholar's Press.

Fenske, N.A. and Lober, C.W. (1986). Structural and functional changes of normal aging skin. *Journal of the American Academy of Dermatology* 15, 571–85.

Fernández, L., Miró, E., Cano, M. and Buela-Casal, G. (2003). Age-related changes and gender differences in time estimation. *Acta Psychologica* 112, 221–32.

Fieder, M. and Huber, S. (2007). Parental age difference and off-spring count in humans. *Biology Letters* 22, 689–91.

Filene, P.G. (1981). *Men in the Middle*. Englewood Cliffs, New Jersey: Prentice-Hall.

Finch, C.E. (2009). The neurobiology of middle-age has arrived. *Neurobiology of aging* 30, 515–20.

Fogarty, M.P. (1975). *Forty to Sixty*. London: Bedford Square Press.

Foley, R.A. and Lee, P.C. (1991). Ecology and energetics of encephalization in hominid evolution. *Philosophical Transactions of the Royal Society of London B: Biological Sciences* 334, 223–31.

Fortunato, L. and Archetti, M. (2010). Evolution of monogamous marriage by maximization of inclusive fitness. *Journal of Evolutionary Biology* 23, 149–56.

Fox, M., Sear, R., Beise, J., Ragsdale, G., Voland, E. and Knapp, L.A. (2010). Grandma plays favourites: X-chromosome relatedness and sex-specific childhood mortality. *Proceedings of the Royal Society B* 277, 567–73.

Freund, A.M. and Ritter, J.O. (2009). Midlife crisis: a debate. *Gerontology* 55, 582–91.

Frisch, R.E. (2002). *Female Fertility and the Body Fat Connection*. Chicago, Illinois: Chicago University Press.

Garcia, J.R., MacKillop, J., Aller, E.L., Merriwether, A.M., Wilson, D.S. and Lum, J.K. (2010). Associations between dopamine D4 receptor gene variation with both infidelity and sexual promiscuity. *PLoS One* 5, e14162.

Garcia, L.T. and Markey, C. (2007). Matching in sexual experience for married, cohabitating, and dating couples. *Journal of Sex Research* 44, 250–5.

Gavrilova, N.S., Gavrilov, L.A., Semyonova, V.G. and Evdokushkina, G.N. (2004). Does exceptional human longevity come with a high cost of infertility? Testing the evolutionary theories of aging. *Annals of the New York Academy of Science* 1019, 513–17.

Genovese, R.G. (1997). *Americans at Midlife*. Westport, Connecticut: Bergin & Garvey.

Gibson, M.A., Mace, R. (2007). Polygyny, reproductive success and child health in rural Ethiopia: why marry a married man? *Journal of Biosocial Science* 39, 287–300.

Gilbert, P. and Allan, S. (1998). The role of defeat and entrapment (arrested flight) in depression: an exploration of an evolutionary view. *Psychological Medicine* 28, 585–98.

Glenn, N. (2009). Is the apparent U-shape of well-being over the life course a result of inappropriate use of control variables? A commentary on Blanchflower and Oswald. *Social Science and Medicine* 69, 481–5.

Gofrit, O.N. (2006). The evolutionary role of erectile dysfunction. *Medical Hypotheses* 67, 1245–9.

Goldbacher, E.M., Bromberger, J. and Matthews, K.A. (2009). Lifetime history of major depression predicts the development of the metabolic syndrome in middle-aged women. *Psychosomatic Medicine* 71, 266–72.

Gould, S.J. (2002). *The Structure of Evolutionary Theory*. Cambridge, Massachusetts: Harvard University Press.

Guillemard, A.M. (1972) *La Retraite – une Morte Sociale*. Paris: La Haye: Mouton.

Gunn, D.A. et al. (2009). Why some women look young for their age. *PLoS One* 4, e8021.

Gunstad, J., Cohen, R.A., Paul, R.H., Luyster, F.S. and Gordon, E. (2006). Age effects in time estimation: relationship to frontal brain morphometry. *Journal of Integrative Neuroscience* 5, 75–87.

Gustafson, D.R. and others (2009). Adiposity indicators and dementia over 32 years in Sweden. *Neurology* 73, 1559–66.

Guyuron, B., Rowe, D.J., Weinfeld, A.B., Eshraghi, Y., Fathi, A. and Iamphongsai, S. (2009). Factors contributing to the facial aging of identical twins. *Plastic and Reconstructive Surgery* 123, 1321–31.

Hager, L.D. (1997). *Women in Human Evolution.* London: Routledge.

Hammock, E.A. and Young, L.J. (2006). Oxytocin, vasopressin and pair bonding: implications for autism. *Philosophical Transactions of the Royal Society of London B: Biological Sciences* 361, 2187–98.

Hampson, S.E., Goldberg, L.R., Vogt, T.M. and Dubanoski, J.P. (2006). Forty years on: teachers' assessments of children's personality traits predict self-reported health behaviors and outcomes at midlife. *Health Psychology* 25, 57–64.

Hancock, P.A. (2010). The effect of age and sex on the perception of time in life. *American Journal of Psychology* 123, 1–13.

Hancock, P.A. and Rausch, R. (2010). The effects of sex, age, and interval duration on the perception of time. *Acta Psychologica* 133, 170–9.

Harlow, B.L. and Signorello, L.B. (2000). Factors associated with early menopause. *Maturitas* 35, 3–9.

Harris, M.B. (1994). Growing Old Gracefully: Age Concealment and Gender. *Journal of Gerontology* 49, 149–58.

Hartmann, U., Philippsohn, S., Heiser, K. and Rüffer-Hesse, C. (2004). Low sexual desire in midlife and older women: personality factors, psychosocial development, present sexuality. *Menopause* 11, 726–40.

Haub, C. (1995). How many people have ever lived on earth? *Population Today* 23, 4–5.

Hawkes, K. (2003). Grandmothers and the evolution of human longevity. *American Journal of Human Biology* 15, 380–400.

Hayes, A.F. (1995). Age preferences for same- and opposite-sex partners. *Journal of Social Psychology* 135, 125–33.

Heckhausen, J. and Schulz, R. (1995). A life-span theory of control. *Psychological Review* 102, 284–304.

Helson, R. and Moane, G. (1987). Personality change in women from college to midlife. *Journal of Personality and Social Psychology* 53, 176–86.

Herbst, J.H., McCrae, R.R., Costa, P.T. Jr, Feaganes, J.R. and Siegler, I.C. (2000). Self-perceptions of stability and change in personality at midlife: the UNC Alumni Heart Study. *Assessment* 7, 379–88.

Heys, K.R., Friedrich, M.G. and Truscott, R.J. (2007). Presbyopia and heat: changes associated with aging of the human lens suggest a functional role for the small heat shock protein, alpha-crystallin, in maintaining lens flexibility. *Aging Cell* 6, 807–15.

Hill, K. and Hurtado, A.M. (1991). The evolution of premature reproductive senescence and menopause in human females: an evaluation of the 'grandmother' hypothesis. *Human Nature* 2, 313–50.

Hill, K. and Hurtado, A.M. (2009). Cooperative breeding in South American hunter-gatherers. School of Human Evolution and Social Change. *Proceedings. Biological Sciences* 276, 3863–70.

Hill, K., Hurtado, A.M. and Walker, R.S. (2007). High adult mortality among Hiwi hunter-gatherers: implications for human evolution. *Journal of Human Evolution* 52, 443–54.

Hill, S.E. and Buss, D.M. (2008). The mere presence of opposite-sex others on judgments of sexual and romantic desirability: opposite effects for men and women. *Personality and Social Psychology Bulletin* 34, 635–47.

Huang, L., Sauve, R., Birkett, N., Fergusson, D. and van Walraven, C. (2008). Maternal age and risk of stillbirth: a systematic review. *Canadian Medical Association Journal* 178, 165–72.

Hultén, M.A., Patel, S., Jonasson, J. and Iwarsson, E. (2010). On the origin of the maternal age effect in trisomy 21 Down syndrome: the Oocyte Mosaicism Selection model. *Reproduction* 139, 1–9.

Huppert, F.A. and Baylis, N. (2004). Well-being: towards an integration of psychology, neurobiology and social science. *Philosophical Transactions of the Royal Society of London B: Biological Sciences* 359, 1447–51.

Huppert, F.A., Abbott, R.A., Ploubidis, G.B., Richards, M. and Kuh, D. (2010). Parental practices predict psychological well-being in midlife: life-course associations among women in the 1946 British birth cohort. *Psychological Medicine* 40, 1507–18.

Hurwitz, J.M. and Santoro, N. (2004). Inhibins, activins, and follistatin in the aging female and male. *Seminars in Reproductive Medicine* 22, 209–17.

Imokawa, G. (2009). Mechanism of UVB-induced wrinkling of the skin: paracrine cytokine linkage between keratinocytes and fibroblasts leading to the stimulation of elastase. *Journal of Investigative Dermatology, Symposium Proceedings* 14, 36–43.

Jaffe, D.H., Eisenbach, Z., Neumark, Y.D. and Manor, O. (2006). Effects of husbands' and wives' education on each other's mortality. *Social Science & Medicine* 62, 2014–23.

James, W.H (1983). Decline in Coital Rates with Spouses' Ages and Duration of Marriage. *Journal of Biosocial Science* 15, 83–7.

Jasienska, G., Nenko, I. and Jasienski, M. (2006). Daughters increase longevity of fathers, but daughters and sons equally reduce longevity of mothers. *American Journal of Human Biology* 18, 422–5.

Jéquier, E. (2002). Leptin signaling, adiposity, and energy balance. *Annals of the New York Academy of Science* 967, 379–88.

Johnstone, R.A. and Cant, M.A. (2010). The evolution of menopause in cetaceans and humans: the role of demography. *Proceedings. Biological Sciences* 277, 3765–71.

Jones, T.M., Balmford, A. and Quinnell, R.J. (2000). Adaptive female choice for middle-aged mates in a lekking sandfly. *Proceedings of the Royal Society of London Series B: Biological Sciences* 267, 681–6.

Joseph, R. (2000). The evolution of sex differences in language, sexuality, and visual-spatial skills. *Archives of Sexual Behavior* 29, 35–66.

Joubert, C.E. (1983). Subjective acceleration of time: death anxiety and sex differences. *Perceptual and Motor Skills* 57, 49–50.

Juul, A. and Skakkebaek, N.E. (2002). Androgens and the ageing male. *Human Reproduction Update* 8, 423–33.

Kalmijn, S., van Boxtel, M.P., Ocké, M., Verschuren, W.M., Kromhout, D. and Launer, L.J. (2004). Dietary intake of fatty acids and fish in relation to cognitive performance at middle age. *Neurology* 62, 275–80.

Kaplan, H.S. and Robson, A.J. (2002). The emergence of humans: the coevolution of intelligence and longevity with intergenerational transfers. *Proceedings of the National Academy of Sciences USA* 99, 10221–6.

Kaplan, H.S., Gurven, M., Winking, J., Hooper, P.L. and Stieglitz, J.

(2010). Learning, menopause, and the human adaptive complex. *Annals of the New York Academy of Science* 1204, 30–42.

Kaplan, H.S., Hill, K., Lancaster, J. and Hurtado, A.M. (2000). A theory of human life history evolution. *Evolutionary Anthropology* 9, 156–85.

Kaplan, H.S., Hooper, P.L. and Gurven, M. (2009). The evolutionary and ecological roots of human social organization. *Philosophical Transactions of the Royal Society of London B: Biological Sciences* 364, 3289–99.

Kaplan, H.S., Lancaster, J.B., Tucker, W.T. and Anderson, K.G. (2002). Evolutionary approach to below replacement fertility. *American Journal of Human Biology* 14, 233–56.

Kauth, M.R. (2006). *Handbook of the Evolution of Human Sexuality.* Binghamton, New York: Haworth Press.

Kemkes, A. (2008). Is perceived childlessness a cue for stereotyping? Evolutionary aspects of a social phenomenon. *Biodemography and Social Biology* 54, 33–46.

Kennedy, K.M. and Raz, N. (2009). Aging white matter and cognition: differential effects of regional variations in diffusion properties on memory, executive functions, and speed. *Neuropsychologia* 47, 916–27.

Kenrick, R.C. and Keefe, D.T. (1992). Age preferences in mates reflect sex differences in human reproductive strategies. *Behavioral and Brain Sciences* 15, 75–113.

King, D.E., Cummings, D. and Whetstone, L. (2005). Attendance at religious services and subsequent mental health in midlife women. *International Journal of Psychiatry in Medicine* 35, 287–97.

Kirkwood, T.B. (2008). Understanding ageing from an evolutionary perspective. *Journal of Internal Medicine* 263, 117–27.

Kirkwood, T.B. and Holliday, R. (1979). The evolution of ageing and longevity. *Philosophical Transactions of the Royal Society of London B: Biological Sciences* 205, 531–46.

Kleiman, D.G. (1977). Monogamy in mammals. *Quarterly Review of Biology* 52, 39–69.

Kopelman, P.G. (1997). The effects of weight loss treatments on upper and lower body fat. *International Journal of Obesity and Related Metabolic Disorders* 21, 619–25.

Kruger, A. (1994). The midlife transition: crisis or chimera. *Psychological Reports* 75, 1299–1305.

Kuhle, B.X. (2007). An evolutionary perspective on the origin and ontogeny of menopause. *Maturitas* 57, 329–37.

Labuda, D., Lefebvre, J.F., Nadeau, P. and Roy-Gagnon, M.H. (2010). Female-to-male breeding ratio in modern humans – an analysis based on historical recombinations. *American Journal of Human Genetics* 86, 353–63.

Lahdenperä, M., Lummaa, V., Helle, S., Tremblay, M. and Russell, A.F. (2004). Fitness benefits of prolonged post-reproductive lifespan in women. *Nature* 428, 178–81.

Larke, A. and Crews, D.E. (2006). Parental investment, late reproduction, and increased reserve capacity are associated with longevity in humans. *Journal of Physiological Anthropology*, 25, 119–31.

Larsen, C.S. (2000). *Skeletons in Our Closet: Reading Our Past Through Bioarchaeology*. Princeton, New Jersey: Princeton University Press.

Lassek, W.D. and Gaulin, S.J. (2006). Changes in body fat distribution in relation to parity in American women: a covert form of maternal depletion. *American Journal of Physical Anthropology* 131, 295–302.

Lee, R.B (1969). !Kung bushman subsistence: an input-output analysis. In Vayda, A., ed. *Ecological Studies in Cultural Anthropology*. Garden City, New York: Natural History Press.

Lehrer, E. and Nerlove, M. (1981). The labor supply and fertility behavior of married women: a three-period model. *Research in Population Economics* 3, 123–45.

Lemlich, R. (1975). Subjective acceleration of time with aging. *Perceptual and Motor Skills* 41, 235–8.

León, M.S. and others (2008). Neanderthal brain size at birth provides insights into the evolution of human life history. *Proceedings of the National Academy of Sciences USA* 105, 13764–5.

Leonard, W.R. and Robertson, M.L. (1997). Comparative primate energetics and hominid evolution. *American Journal of Physiological Anthropology* 102, 265–81.

Levine, S.B. (1998). *Sexuality in Midlife*. New York, New York: Plenum Press.

Levinson, D.J., Darrow, C.N., Klein, E.B., Levinson, M.H. and McKee, B. (1978). *The Seasons of a Man's Life*. New York, New York: Knopf.

Lev-Ran, A. (2001). Human obesity: an evolutionary approach to understanding our bulging waistline. *Diabetes/Metabolism Research and Reviews* 17, 347–62.

Levy, B. (2001) Eradication of ageism requires addressing the enemy within. *The Gerontologist* 41, 5.

Lewis, K. (1999). Human longevity: an evolutionary approach. *Mechanisms of Ageing and Development*, 109, 43–51.

Lindau, S.T. and Gavrilova, N. (2010). Sex, health, and years of sexually active life gained due to good health: evidence from two US population based cross sectional surveys of ageing. *British Medical Journal* 340, c810.

Lindau, S.T., Schumm, L.P., Laumann, E.O., Levinson, W., O'Muircheartaigh, C.A. and Waite, L.J. (2007). A study of sexuality and health among older adults in the United States. *New England Journal of Medicine* 357, 762–74.

Ljubuncic, P. and Reznick, A.Z. (2009). The evolutionary theories of aging revisited – a mini-review. *Gerontology*, 55, 205–16.

Longo, V.D., Mitteldorf, J. and Skulachev, V.P. (2005). Programmed and altruistic ageing. *Nature Reviews Genetics* 6, 86672.

Lookingbill, D.P., Demers, L.M., Wang, C., Leung, A., Rittmaster, R.S. and Santen, R.J. (1991). Clinical and biochemical parameters of androgen action in normal healthy Caucasian versus Chinese subjects. *Journal of Clinical Endocrinology and Metabolism* 72, 1242–8.

Lyons, M.J. et al. (2009). Genes determine stability and the environment determines change in cognitive ability during 35 years of adulthood. *Psychological Science*, 20, 1146–52.

Malaspina, D. and others. (2005). Paternal age and intelligence:

implications for age-related genomic changes in male germ cells. *Psychiatric Genetics* 15, 117–25.

Malaspina, D., Perrin, M., Kleinhaus, K.R., Opler, M., Harlap, S. (2008). Growth and schizophrenia: aetiology, epidemiology and epigenetics. *Novartis Foundation Symposia* 289, 196–203.

Martin, R.D. (2007). The evolution of human reproduction: a primatological perspective. *American Journal of Physical Anthropology* supplement 45, 59-84.

Mattila, V. (1987). Onset of functional psychoses in later middle age. Social-psychiatric considerations. *Acta Psychiatrica Scandinavica* 76, 293–302.

McAdams, D.P. and Olson B.D. (2010). Personality development: continuity and change over the life course. *Annual Review of Psychology* 61, 517–42.

Megarry, T. (2005). *Society in Prehistory*. Basingstoke: Macmillan Press.

Miller, G.F. (2000). *The Mating Mind*. New York, New York: Anchor Books.

Mojtabai, R. and Olfson, M. (2004). Major depression in community-dwelling middle-aged and older adults: prevalence and 2- and 4-year follow-up symptoms. *Psychological Medicine* 34, 623–34.

Muller, M.N., Thompson, M.E. and Wrangham, R.W. (2006). Male chimpanzees prefer mating with old females. *Current Biology* 16, 2234–8.

Murphy, C., Wetter, S., Morgan, C.D., Ellison, D.W. and Geisler, M.W. (1998). Age effects on central nervous system activity reflected in the olfactory event-related potential. Evidence for decline in middle age. *Annals of the New York Academy of Science* 855, 598–607.

Murstein, B.I. and Christy, P. (1976). Physical attractiveness and marriage adjustment in middle-aged couples. *Journal of Personality and Social Psychology*, 34, 537–42.

Neu, P., Bajbouj, M., Schilling, A., Godemann, F., Berman, R.M. and Schlattmann, P. (2005). Cognitive function over the treatment course of depression in middle-aged patients: correlation with brain MRI signal hyperintensities. *Journal of Psychiatric Research* 39, 129–35.

Nitardy, F.W. (1943). Apparent time acceleration with age of the individual. *Science* 98, 110.

North, R.J., Holahan, C.J., Moos, R.H. and Cronkite, R.C. (2008). Family support, family income, and happiness: a 10-year perspective. *Journal of Family Psychology* 22, 47583.

Nyklícek, I., Louwman, W.J., Van Nierop, P.W., Wijnands, C.J., Coebergh, J.W. and Pop, V.J. (2003). Depression and the lower risk for breast cancer development in middle-aged women: a prospective study. *Psychological Medicine* 33, 111117.

O'Connor, D.B. and others (2009). Cortisol awakening rise in middle-aged women in relation to psychological stress. *Psychoneuroendocrinology* 34, 1486–94.

Ostovich, J.M. and Rozin, P. (2004). Body image across three generations of Americans: inter-family correlations, gender differences, and generation differences. *Eating and Weight Disorders* 9, 186–93.

Oswald, A.J (1997). Happiness and economic performance. *Economic Journal* 107, 1815–31.

Partridge, L. (201). Evolutionary theories of ageing applied to long-lived organisms. *Experimental Gerontology* 36, 641–50.

Pavard, S.E., Metcalf, C.J. and Heyer, E. (2008). Senescence of reproduction may explain adaptive menopause in humans: a test of the 'mother' hypothesis. *American Journal of Physical Anthropology* 136, 194–203.

Penn, D.J. and Smith, K.R. (2007). Differential fitness costs of reproduction between the sexes. *Proceedings of the National Academy of Sciences USA* 104, 553–8.

Peters, J. and Daum, I (2008). Differential effects of normal aging on recollection of concrete and abstract words. *Neuropsychology* 22, 255–61.

Phillips, L.H. and Allen, R. (2004). Adult aging and the perceived intensity of emotions in faces and stories. *Aging Clinical and Experimental Research* 16, 190–1.

Power, M.L. and Schulkin, J. (2008). Sex differences in fat storage, fat metabolism, and the health risks from obesity: possible evolutionary origins. *British Journal of Nutrition* 99, 931–40.

Power, M.L. and Schulkin, J. (2009). *The Evolution of Obesity*. Baltimore: Johns Hopkins University Press.

Prapas, N., Kalogiannidis, I., Prapas, I., Xiromeritis, P., Karagiannidis, A. and Makedos, G. (2006). Twin gestation in older women: antepartum, intrapartum complications, and perinatal outcomes. *Archives of Gynecology and Obstetrics* 273, 293–7.

Ramsawh, H.J., Raffa, S.D., Edelen, M.O., Rende, R. and Keller, M.B. (2009). Anxiety in middle adulthood: effects of age and time on the 14-year course of panic disorder, social phobia and generalized anxiety disorder. *Psychological Medicine* 39, 615–24.

Ransohoff, R.M. (1990). *Venus After Forty*. Far Hills, New Jersey: New Horizon Press.

Rao, K.V. and Demaris, A. (1995). Coital frequency among married and cohabiting couples in the United States. *Journal of Biosocial Science* 27, 135–50.

Rashidi, A. and Shanley, D. (2009). Evolution of the menopause: life histories and mechanisms. *Menopause International* 15, 26–30.

Reichman, N.E. and Pagnini, D.L. (1997). Maternal age and birth outcomes: data from New Jersey. *Family Planning Perspectives* 29, 268–72.

Reid, J. and Hardy, M. (1999). Multiple roles and well-being among midlife women: testing role strain and role enhancement theories. *The Journals of Gerontology. Series B, Psychological Sciences and Social Sciences* 54, S329–38.

Resnick, S.M., Lamar, M. and Driscoll, I. (2007). Vulnerability of the orbitofrontal cortex to age-associated structural and functional brain changes. *Annals of the New York Academy of Science* 1121, 562–75.

Riddle, J. (1992). *Contraception and Abortion from the Ancient World to the Renaissance*. Cambridge, Massachusetts: Harvard University Press.

Riis, J.L., Chong, H., Ryan, K.K., Wolk, D.A., Rentz, D.M., Holcomb, P.J. and Daffner, K.R. (2008). Compensatory neural activity distinguishes different patterns of normal cognitive aging. *NeuroImage* 39, 441–54.

Ritz-Timme. S. and others (2000). Age estimation: the state of the art in relation to the specific demands of forensic practise. *International Journal of Legal Medicine* 113, 129–36.

Robson, S.L. and Wood, B. (2008). Hominin life history: reconstruction and evolution. *Journal of Anatomy*, 212, 394–425.

Rose, M.R. (1991). *Evolutionary Biology of Aging*. New York, New York: Oxford University Press.

Rossi, A.S. (1994). *Sexuality across the Life Course*. Chicago, Illinois: Chicago University Press.

Ryff, C.D. (1989). In the eye of the beholder: views of psychological well-being among middle-aged and older adults. *Psychology and Aging* 4, 195–201.

Sauvain-Dugerdil, C., Leridon, H., Mascie-Taylor, N., eds. (2006). *Human Clocks: The Bio-cultural Meanings of Age*. Bern: Peter Lang.

Schmidt, P.J., Murphy, J.H., Haq, N., Rubinow, D.R. and Danaceau, M.A. (2004). Stressful life events, personal losses, and perimenopause-related depression. *Archives of Women's Mental Health* 7, 19–26.

Schmitt, D.P. (2005). Fundamentals of human mating strategies. In Buss, D.M., ed. *Handbook of Evolutionary Psychology*. Hoboken, New Jersey: John Wiley and Sons.

Schmitt, D.P. (2005). Is short-term mating the maladaptive result of insecure attachment? A test of competing evolutionary perspectives. *Personality and Social Psychology Bulletin*, 31, 747–68.

Schnarch, D. (1997). *Passionate Marriage*. New York, New York: Henry Holt.

Segal, S.J. and Mastroianni, L. (2003). *Hormone Use in Menopause and Male Andropause*. New York, New York: Oxford University Press.

Sherman, C.A., Harvey, S.M. and Noell, J. (2005). 'Are they still having sex?' STIs and unintended pregnancy among mid-life women. *Journal of Women & Aging* 17, 41–55.

Singh-Manoux, A., Richards, M. and Marmot, M. (2003). Leisure activities and cognitive function in middle age: evidence from the Whitehall II study. *Journal of Epidemiology and Community Health* 57, 907–13.

Skultety, K.M. and Krauss Whitbourne, S. (2004). Gender differences in identity processes and self-esteem in middle and later adulthood. *Journal of Women and Aging* 16, 175–88.

Stanford, J.L., Hartge, P., Brinton, L.A., Hoover, R.N. and Brookmeyer, R. (1987). Factors influencing the age at natural menopause. *Journal of Chronic Diseases* 40, 995–1002.

Sternberg, R.J. and Grigorenko, E.L. (2004). Intelligence and culture: how culture shapes what intelligence means, and the implications for a science of well-being. *Philosophical Transactions of the Royal Society of London B: Biological Sciences* 359, 1427–34.

Stevens, J., Katz, E.G. and Huxley, R.R. (2010). Associations between gender, age and waist circumference. *European Journal of Clinical Nutrition* 64, 6–15.

Stevens, J.C. (1992). Aging and spatial acuity of touch. *Journal of Gerontology* 47, 35–40.

Stewart, A.J. and Vandewater, E.A. (1999). 'If I had it to do over again . . .': midlife review, midcourse corrections, and women's well-being in midlife. *Journal of Personality and Social Psychology* 76, 270–83.

Strassmann, B.I. (1999). Menstrual cycling and breast cancer: an evolutionary perspective. *Journal of Women's Health* 8, 193–202.

Strehler, B.L. (1979). Polygamy and the evolution of human longevity. *Mechanisms of Ageing and Development* 9, 369–79.

Taylor, T. (1996). *The Prehistory of Sex.* London: Fourth Estate.

Thompson, E.H. (1994). *Older Men's Lives.* Thousand Oaks, California: Sage.

Tishkoff, S.A. et al. (2001). Haplotype diversity and linkage disequilibrium at human G6PD: recent origin of alleles that confer malarial resistance. *Science* 293, 455–62.

Tobin, D.J. and Paus, R. (2001). Graying: gerontobiology of the hair follicle pigmentary unit. *Experimental Gerontology* 36, 29–54.

Tobin, D.J., Hordinsky, M. and Bernard, B.A. (2005). Hair pigmentation: a research update. *Journal of Investigative Dermatology, Symposium Proceedings* 10, 275–9.

Tse, P.U., Intriligator, J., Rivest, J. and Cavanagh, P. (2004). Attention and the subjective expansion of time. *Perception and Psychophysics* 66, 1171–89.

Tuljapurkar, S.D., Puleston, C.O. and Gurven, M.D. (2007). Why men matter: mating patterns drive evolution of human lifespan. *PLoS One* 2, e785.

Uitto, J. (2008). The role of elastin and collagen in cutaneous aging: intrinsic aging versus photoexposure. *Journal of Drugs in Dermatology* 7, supplement 2, s12–16.

Vaillant, G.E, Bond, M. and Vaillant, C.O. (1986). An empirically validated hierarchy of defense mechansims. *Archives of General Psychiatry* 43, 786–94.

Walker, J.L. (1977). Time estimation and total subjective time. *Perceptual and Motor Skills* 44, 527–32.

Walker, R. et al. (2006). Growth rates and life histories in twenty-two small-scale societies. *American Journal of Human Biology* 18, 295–311.

Wang, M. H. and vom Saal, F.S. (2000). Maternal age and traits in offspring. *Nature* 407, 469–70.

Ward, E.J., Parsons, K., Holmes, E.E., Balcomb, K.C. and Ford, J.K. (2009). The role of menopause and reproductive senescence in a long-lived social mammal. *Frontiers in Zoology* 6, 4.

Waters, D.J., Shen, S. and Glickman, L.T. (2000). Life expectancy, antagonistic pleiotropy, and the testis of dogs and men. *Prostate* 43, 272–7.

Weisfeld, G.E. and Weisfeld, C.C. (2002). Marriage: an evolutionary perspective. *Neuroendocrinology Letters*, 23 supplement 4, 47–54.

Wells, J.C. and Stock, J.T. (2007). The biology of the colonizing ape. *American Journal of Physical Anthropology* 134 supplement 45, 191–222.

Wells, J.C. (2006). The evolution of human fatness and susceptibility to obesity: an ethological approach. *Biological Reviews of the Cambridge Philosophical Society* 81, 183–205.

Wespes, E. (2002). The ageing penis. *World Journal of Urology* 30, 36–9.

Whelan, E.A., Sandler, D.P., McConnaughey, D.R. and Weinberg, C.R. (1990). Menstrual and reproductive characteristics and age at natural menopause. *Americal Journal of Epidemiology*, 131, 625–32.

Williams, G.C. (1957). Pleiotropy, natural selection and the evolution of senescence. *Evolution* 11, 398–411.

Willis, S.L. and Reid, J.D., eds. (1999). *Life in the Middle*. San Diego, California: Academic Press.

Wilson, E.O. (1975). *Sociobiology: The New Synthesis*. Cambridge, Massachusetts: Harvard University Press.

Winking, J., Kaplan, H., Gurven, M. and Rucas, S. (2007). Why do men marry and why do they stray? *Proceedings. Biological Sciences* 274, 1643–9.

Wittmann, M. and Lehnhoff, S. (2005). Age effects in perception of time. *Psychological Reports* 97, 921–35.

Wodinsky, J. (1977). Hormonal Inhibition of Feeding and Death in Octopus: Control by Optic Gland Secretion. *Science* 148, 948–51.

Wood, B.J. (2000). Investigating human evolutionary history. *Journal of Anatomy* 197, 3–17.

Wood, J.M. and others (2009). Senile hair graying: H_2O_2-mediated oxidative stress affects human hair color by blunting methionine sulfoxide repair. *FASEB Journal* 23, 2065–75.

Yarrow, A.L. (1991). *Latecomers: Children of Parents over 35*. Old Tappan, New Jersey: Free Press.

Yassin, A.A. and Saad, F. (2008). Testosterone and erectile dysfunction. *Journal of Andrology* 29, 593–604.

Zafon, C. (2007). Oscillations in total body fat content through life: an evolutionary perspective. *Obesity Reviews* 8, 525–30.

Zamboni, G., Gozzi, M., Krueger, F., Duhamel, J.R., Sirigu, A. and Grafman, J. (2009). Individualism, conservatism, and radicalism as criteria for processing political beliefs: a parametric fMRI study. *Social Neuroscience* 4, 367–83.

Zerjal, T. et al. (2003). The genetic legacy of the Mongols. *American Journal of Human Genetics* 72, 717–21.

Zhu, J.L., Vestergaard, M., Madsen, K.M. and Olsen, J. (2008). Paternal age and mortality in children. *European Journal of Epidemiology* 23, 443–7.

Index

and parental depression,
158–9, 163
parenting, 51–60, 209–14,
243–5
psychological development,
130
relationship between
upbringing and depression,
159
reluctant to leave home, 254–5
chimpanzees
age of adulthood, 54
and ageing, 24
hair greying, 70
life expectancy, 31, 41
sex life, 223–4
and social hierarchies, 160
cholesterol, 86, 171
cognition *see* brain
collagen, 12–13, 64–6
conservatism, and age, 134–8
control, importance of sense of,
130–4
cortisol, 161–2
cosmetics, 65–6
culture
swift change of, 112
transmission of, 56–9, 174

Darwin, Charles, 18
death
death-anxiety and time
perception, 114–15
immortal animals, 25–6
middle-age preoccupation
with, 23
reasons for inevitability, 26–35
see also life expectancy
deer, 160, 186

depression
decline in middle age, 156–8
and empty-nest syndrome, 249
and menopause, 205
overview, 154, 155–65
physical and mental effects,
156
physiological causes, 161–2
possible causes, 158–65
reactive and clinical defined,
155–6
dermis, 63
development
long period of human, 51–2
and middle age, 14–17
post-natal vs pre-natal, 15–16
diabetes, 85–6, 220
diet
and advent of agriculture,
44–5, 86–7
human food-gathering, 53–5
dik dik, 266
diseases
and advent of agriculture,
45–6
caused by parental age, 240–2
in hunter-gatherer society, 45
and premature ageing, 34–5
disposable soma theory, 29–30
divorce, 260–2
DNA
and ageing, 32–3, 35
and middle age, 12–14
dogmatism, 173–5
dopamine receptors, 107
Down's syndrome, 236–8, 240
Dubin, Al, 61
dwarfism, 240

rates of marriage, 260–1
reproductive lifespans, 50
sense of control, 132–3
sense of identity, 134
sense of taste, 98
time perception, 122
genes
and ageing, 28–9, 32–5
and cognitive ability, 172–3
and emotions, 144–5
and middle age, 11–21
reuse for different purposes,
15
Genghis Khan, 267
gibbons, 268
Gilbert, P., 154
glands, 63
gonadotrophins, 200–1, 203,
218–19
gorillas, 41, 70, 268
grandparents, 210–13

Hadra people, 44
hair
distribution, 16, 71–2
greying, 69–71
hair follicles, 63, 71
hamsters, 30, 120
hand grip, 82
hands, 68
happiness
and achievement, 145
and the brain, 145–7
as distinct from well-being,
147
and genetics, 144–5
and middle age, 128, 139–47,
151–3
and old age, 141–2, 143

health
and cognitive ability, 170–1
and depression, 162–3
see also mental health
hearing, 97–8
heart disease
and body fat, 85–6
and cognitive ability, 171
and depression, 162
and erectile dysfunction, 220
evolutionary reason for, 87–8
and menopause, 204–5
heat tolerance, 68
heredity see genes
Hobbes, Thomas, 37
homemaking, and depression, 160
homosexual relationships, 223,
264
Hormone Replacement Therapy
(HRT), 213–14
hormones, sex, 187–8
hot flushes, 204–5
humans: special nature of species,
49–52
hunter-gatherers
life expectancy, 41–7
modern, 42, 54, 55

identity, sense of, 133–4
Iglesias, Julio, 239
infidelity, 272–5
inflexibility, 173–5
inhibin, 203–4
intelligence see brain
introspection, middle-age, 59,
175–7

jealousy, sexual, 275, 277
Jung, Karl, 127

Keep in touch with
Portobello Books:

Visit portobellobooks.com to discover more.

Portobello